Creative Propagation

A GROWER'S GUIDE

Peter Thompson

I think the true gardener is a lover of his flowers, not a critic of them. I think the true gardener is the reverent servant of Nature, not her truculent, wife-beating master. I think the true gardener, the older he grows should more and more develop a humble, grateful and uncertain spirit, cocksure of nothing except the universality of beauty.

Reginald Farrer, *In A Yorkshire Garden*

D1042431

TIMBER PRESS
Portland, Oregon

First published in 1989
This edition first published 1992
Reprinted 1993, 1994

Typeset by Paston Press, Loddon, Norfolk
and printed in Great Britain by Biddles Ltd, Guildford

Published by
B. T. Batsford Ltd
4 Fitzhardinge Street
London W1H 0AH

A catalogue record for this book is available from the British Library

ISBN 0 71347118 2

Published in North America by
Timber Press, Inc.
The Haseltine Building
133 S.W. Second Ave., Suite 450
Portland, Oregon 97204-3527, U.S.A.

ISBN 0-88192-251-X

Contents

CHAPTER ONE

Prelude to Propagation

It's a pretty good bet that all who pick up this book will, sooner or later, be infuriated by something they read in it; some description of a technique, or an opinion expressed, will produce a snort of derision or the conviction that the author must be an idiot. Propagation affects most of us like that. We all have our pet ways of doing things. Some of us think our methods work because of the care we are prepared to take; some of us believe we are successful because we don't go to the extremes of unnecessary expense or the time-consuming fussiness of our more punctilious friends. The mistake we all incline to make is to suppose that the difference between successfully and unsuccessfully growing seedlings, or persuading cuttings to produce roots, just depends on knowing facts. It doesn't; it depends much more on getting to know plants. It depends on knowing them well enough to recognise when they are suffering stress before the symptoms become obvious. It depends on being able, almost instinctively, to see to their needs so that they are able to do all the things they are capable of. Nobody ever yet rooted a cutting or germinated a seed; those are things that only plants can do and the propagator's job is to find the best ways of letting them get on with it.

There is a temptation to regard propagation as a horticultural technique, because this approach provides a hope that success depends on acquiring technical skills, and using appropriate equipment and other resources in a carefully specified way for any particular plant. If this were so then any book on propagation would only have to set out what should be done—much like the directions, on the outside of the pack, for assembling a fold-flat garden frame—and, so long as one knew how to use a screwdriver, and carefully followed the instructions, all would be well. This approach with its rule of thumb methods can be very effective, and provides the foundation on which commercial plant production is now becoming more and more dependent. But, it brings with it the impression that propagation is something that human beings have invented, and has very little to do with the way plants grow naturally. It can suggest that it is a more or less artificial process in which the cost and sophistication of the facilities used are the most important elements. This impression is not reduced by a visit to any well-stocked garden centre. There the display will include beguiling arrays of equipment, ranging from pre-sown propagation kits, specially designed containers, artfully formulated composts, electrically heated propagators and automatic watering systems—all arranged to convey a promise of success, ensured by spending money. Unsurprisingly, many people reach the

1

intended conclusion: that successful propagation depends on elaborate equipment, considerable expenditure and a variety of gadgets.

This is not so. Any one of us can kill plants with equal ease, when the conditions we provide for them are not to their liking, whether we use expensive equipment or almost none at all. On the other hand, if the right things are done at the right times, seeds will germinate, cuttings will produce roots and divisions flourish using very simple methods; methods based on those which gardeners have used for centuries, effectively, economically and productively. Nowadays, the ways we employ these methods have been simplified very agreeably by the invention of plastic films, and the availability of materials with which to make composts for sowing seeds, or supporting cuttings. No longer must we depend on the garden boy to syringe cuttings in a sun-frame seven times a day with water, or seek the advice of a formidable head gardener whose lifetime's memory of recipes enables him to concoct a series of magical mixes in which to grow any kind of plant.

The essential thing to recognise is that plant propagation depends on the natural responses which enable plants to survive in wild communities. As an extension to this, it lies in recognising that plants which live in different parts of the world can do so only if they are able to develop the distinctly different responses and capabilities which enable them to cope with the dangers, and take advantages of the opportunities, typical of the places where they live. This has now become a very familiar idea: it is known as natural selection. Whatever our imaginations may conjure up for us, there is no such thing as a typical plant, and no Utopian combination of wet and warmth, light and fertility to which all plants will respond contentedly. Some are adversely affected by high temperatures, some by low. Some rot dismally at the faintest suggestion of being too wet, but endure severe drought without harm. Others wilt and shrivel away at the first sign of a heat wave. Cuttings taken from a plant at one time of the year will produce roots rapidly and abundantly, at another season they will die without producing a single root. Seeds of some kinds of annuals germinate without restraint as soon as they are sown and watered; seeds of another will lie inert and brown and sometimes years will pass before seedlings emerge from them. Recognising these differences by relating them to the natural conditions in which plants survive, and taking them into account constructively, is a much surer and more economical way of propagating plants than relying on sophisticated and expensive equipment.

Propagation is looked on by many people as a rather advanced skill, one which distinguishes the dedicated gardener from the mere dabbler, and certainly there are some plants whose propagation calls for great skill and sensitivity. On the other hand, there are huge numbers of plants—the great majority of those we grow in our gardens—which are not at all difficult to propagate successfully, using very simple methods, which can easily be used by anyone willing to have a go. Plants which we have raised ourselves provide a uniquely personal sense of affinity with them and the place where they grow, which has a great deal to do with the pleasure that our gardens give us, and which the mass-produced products from a garden centre can never bring with them.

But there are other ways in which gardeners profit by producing their own plants. These more definable, practical and useful benefits appeal directly to our material hopes rather than our spiritual needs. They can save a deal of money, especially when masses of plants are needed, perhaps for ground cover, or to plant the large bold groups which look so good, but are so inhibitingly expensive to buy. We may be able to use opportunities, offered by friendly neighbours or generous gardening acquaintances, to propagate some of the numerous plants which are not very easy to find for sale commercially, and nowadays this covers a surprisingly large number of attractive, easily grown things. Many of these are not particularly rare or distinguished, but thoroughly good garden plants which fit comfortably into our gardens, and will flourish with very little special attention or care. It also becomes possible to multiply plants which are really unusual or rare, or perhaps to reproduce special forms of seedlings, or sports which turn up in gardens. These events happen much more often than many people realise—a change of colour, a doubling of a single flower, an alteration of leaf shape or plant form, the appearance of a shoot which produces variegated foliage—all are worth looking out for, and, when they are found, can be reproduced by propagating them to produce a new and unique variety of plant.

During the last few years the ways that nurseries produce plants commercially have changed fundamentally. New methods have been introduced, and the scale of production greatly increased; but changes in the ways plants are marketed and the criteria by which retail sales are assessed have had even greater effects. It has become commonplace for plants to be propagated in one part of the world and sold somewhere far-distant: very often the places where plants are propagated, and the places where they are later sold and grown, have totally different climates, and may even be in different hemispheres. Almost all plants are sold in containers, and there are financial benefits to be gained by encouraging them to grow rapidly so that they produce a saleable plant quickly. The variety of plants generally available has been reduced to exclude those which do not conform easily to production schedules, or which fail to display themselves in ways which encourage impulse buying when they are put up for sale. These changes may be very significant to horticulturalists and marketing organisations which make their living by producing plants for sale economically: they have little relevance to gardeners trying to produce plants for their own gardens. This book is not a manual of nursery stock production, but a source of information for amateur gardeners. Its aim is to describe methods which fit their needs, whether these are recent innovations or ancient practices blessed by tradition. It aims to cover a wide, though not comprehensive, range of the hardy plants that can be grown in their gardens.

A decision that must be made by anyone who writes a book about propagating plants is whether to base it on 'materials' or 'methods'. Should the chapters be based on different plants or on different ways of propagating plants? More often than not the chapter headings of such books are based on methods, each describes a particular topic, such as making divisions, taking hardwood cuttings, sowing seeds, so that one by one they describe all the alternative ways by which plants can be reproduced. This implies that the reader, or user, picks up the book with the intention of, perhaps, spending a day taking hardwood cuttings, or sowing seeds in a heated greenhouse, and would like to make a list of the plants that could be propagated in the chosen way. But, the whole point of propagating something is to obtain more of a particular plant, and it is for this reason that plants, or plants in various groupings, form the major topics in this book, and are the subjects of the chapter headings. One snag with this approach is that there are very many occasions when plants of very different kinds can be propagated by similar methods. For example, shrubs, trees, alpines, annuals and perennial plants of one kind or another can all be propagated from their seeds, and although some may need special conditions that distinguish them from the others, very many in each category can be propagated successfully in precisely similar ways. In order to avoid repetition of this kind, cross-references are used where necessary to direct attention to descriptions of relevant techniques.

We don't all agree on the way we describe our plants. Something which one of us calls an alpine will be a hardy perennial to another. Where do shrubs end and trees begin? Several of the most familiar annuals grown in Britain are perennials in frost-free climates; sometimes there are difficulties with 'sub-shrubs' and similar ambiguities. Apart from the index, which provides the conventional guide to the layout of the book, a Propagation Table (Chapter Fourteen) is also included in which the generic names of the plants are listed with an indication of the methods and times when they can be propagated. The table is based, inevitably, on Latin names, arranged in alphabetical order; but large numbers of common names, with their Latin equivalents, have been included in the index for easy reference. The table provides a convenient summary of the book, but in particular is set out in a way which would make it easy for anyone who wished to do so to construct a propagation calendar covering their own special needs or interests.

Ideas from two distinctly different sources have been brought together during the production of this book. Some come from a period of nearly 20 years spent at The Royal Botanic Gardens at Kew, trying to find out a little more about the ways that plants adapt and change during the processes of acclimatisation that enable them to survive in different parts of the world. Others flow from attempts to help amateur gardeners understand more about the plants they like to grow during courses at The Garden School, and from the day-to-day problems encountered on a nursery, where a large number of hardy plants were propagated to produce stock for sale. Plant propagation may be an art, a science or a technique, and the word chosen to describe it may vary depending on mood, recent experience or optimistic forecasts of future hopes. Whatever view prevails, success depends on the two twin springs of down-to-earth practicalities and the more academic ideas which help us to understand better why plants behave as they do, and how they are likely to respond to the different conditions which we provide with fond imagination for their benefit.

The courses at The Garden School showed just how much scope there is even for successful and experienced gardeners to discover more about straight-forward and simple ways to propagate plants. Successful propagation isn't an inborn, instinctive skill, though a natural sympathy with plants is very helpful, but rather depends on being shown how to do things, being stimulated to think about problems, and being encouraged to attempt our own interpretations of the ways plants grow and their responses to their surroundings. Thus we learn by experience what works and what is not successful. We also learn not to become stereotyped, and stick too rigidly to methods that become habits by long use; methods which may not be particularly successful, may not really be well suited to our needs or even make good use of the space and facilities at our disposal. We all benefit from an occasional good shake up that makes us really look at what we are up to, and think about the tricks and fancies, lore and mumbo-jumbo which have wormed their way into our potting sheds.

But if you have discovered a way to propagate a particular plant, and one which really does work for you, never, ever, be tempted to allow anyone to persuade you to abandon it just because some expert comes along, sniffs at your method, and insists on telling you how much better you could do it in some other way!

CHAPTER TWO

The Natural Reproduction of Plants

Plants of one kind or another can be found growing all over the world, except for those desolate regions which are so cold or so dry that not even one of the multitude of forms that plants are able to adopt is capable of enduring the conditions. They survive by repeatedly reproducing themselves; one generation succeeding another, some year by year, many over much longer periods extending to decades or even to centuries.

Any species of plant that survives in the wild does so only for so long as it reproduces itself successfully, but this critically important part of the pattern of events that makes up a plant's existence is also a risky and uncertain process. At some stage during its course the offspring have to find a way to separate from their parents, and establish themselves successfully elsewhere. It is easy to underestimate the risks involved. Gardeners are constantly beset by weeds, which emerge in their thousands from seeds buried in the soil, or spring up from minute sub-divisions of their parents carelessly left behind during weeding. When the contents of a packet of seeds are sown in containers or in a seed bed, gardeners expect a high proportion to germinate, and should not, usually, be disappointed. They also know that, with care, almost every one of those seedlings can be grown on until it flowers, or produces a succulent vegetable. Under natural conditions seedlings manage to establish themselves from only a tiny proportion of the seeds that are present in the ground, and very few of these grow into mature plants that in their turn produce more seeds to give rise to the next generation. Generally speaking this is something we are all aware of; but, it is salutary to recall how very heavily the dice are loaded against the survival to maturity of any particular seed or seedling.

Sexual Reproduction in Plants

The numbers of plants in any population can change from one year to another, and sometimes these fluctuations are very large indeed—so, one year, cowslips will be found growing in their thousands where before there was barely a scattered handful. But, broadly speaking, over the years, the number of individuals always remains within particular limits, unless something happens which alters the conditions under which the plants are growing. The stark consequence of this is that, on average, in each succeeding generation only one seed of all those produced by each plant successfully produces a surviving seedling. One tall spike of the purple foxglove may produce well over a quarter of a million seeds in its ranks of capsules, to ensure that one survives, grows up to maturity, flowers and in its turn produces seeds. An oak tree, over a life span of hundreds of years, sheds tonnes of acorns on to the ground beneath it. From this mass, one, on average, will find itself in a place where it can germinate, grow to maturity, and in turn produce the successive crops needed to produce one more tree; two generations, perhaps 400 and more years gone by, and just two trees to show for it.

Plants contend with odds against their survival which are not ten to one or even a thousand to one. More usually they are tens or even hundreds of thousands to one, and a population of wild plants will only persist if it is able to produce so many seeds that some individuals still remain after all the agents of death and decay have completed their work.

Reproduction by seed, being a sexual process, results in the genes, carried in the pollen grains and the ovules, being combined and redistributed amongst numerous offspring each of which contains a largely random, 50:50 assortment of the genes of both parents. This ensures that every plant in a population is an individual with its own particular genetic constitution, and these differences between individuals provide populations of plants with diversity and the means to change over the generations by natural selection, during which plants with qualities which favour survival are more likely to produce offspring. Sexual reproduction by seed may be a very hazardous process, in which the odds against the survival of any individual are almost astronomically long, but it can be vitally important to the long-term survival of the population.

Sexual reproduction has its advantages—survival or extinction may often depend on them—but these advantages tend to be only periodically, and sometimes rarely, significant. They are rather similar to contingency plans, and can provide uniquely effective ways to protect a population from total destruction even though such overwhelming threats may menace them only very occasionally, such as an assault by a new strain of disease to which only a few individuals possess resistance, or a spell of extremely severe weather which destroys all but a handful of the hardiest individuals.

Vegetative Reproduction in Plants

Plants have developed less chancy ways to propagate themselves and these have one great advantage: the offspring remain attached to their parents and continue to benefit from support and shelter during the hazardous period when they are setting up as independent individuals. Plants manage this by reproducing themselves from their own growing parts: short shoots grow longer to develop offsets or runners; subterranean, exploratory rhizomes or tubers are produced; branches bend to the ground and layer themselves to form thickets; a mat of stems and foliage spreading across the surface produces roots as it grows. These are asexual methods of reproduction, and are often referred to as vegetative forms of propagation. They depend on the capacity of stems to form the roots needed to make a new and complete plant, and on roots to produce shoots, complete with leaves and ultimately flowers and fruits.

The new plants produced directly from the tissues of their parent are all of a kind, sharing exactly the same genetic constitution one with another and with the single plant from which they originated. Some plants reproduce vegetatively most successfully; they have developed methods of vegetative reproduction as a conservative way of banking on genetic combinations which have already proved themselves. The runners from one plant of a wild strawberry may rapidly cover several square metres, and eventually form a widely dispersed colony that may run the length of a hedgerow. One small plant of a fescue grass, once installed in a spot which suits it, may go on to produce a mat of grass that covers several hectares. The grass may well find itself growing in competition with bracken, equally effective at colonising huge areas from a single plant, and then the dominant flora of an entire hillside may consist essentially of just two plants. A small forest of trunks of the western red cedar will originate from one ancient mother-plant whose branches, trailing to the ground, formed roots where they touched, and then in ever-widening circles the offspring did the same, gradually losing contact one with another as each became established and the links between them disappeared.

Totipotency

Gardeners long ago observed these natural ways of propagation and adapted them for their own purposes. They removed offsets, ready-rooted, as the simplest possible method of making more. They dug up tubers in the autumn, and replanted them the following spring. They found they could pull plants to pieces and multiply them by division. They cut off shoots and small branches to make fences and, after sticking them in the ground, observed that they could form roots and grow into new plants. They sliced the tops off roots and found that they were replaced by a mass of new shoots emerging in a ring from the top of the decapitated root. They cut pieces from two separate plants, bound them together, and saw that they could unite together and grow as one. Latterly, with great care, they cut away the stems and stripped off all the leaves, even those still within the buds, until they were left with a tiny assemblage of cells, the 'growing point' or meristem of the plant, and found that this too could be kept alive, encouraged to grow, and would in time reproduce a plant exactly like that from which it was removed. Finally, abandoning their guise as gardeners altogether, they donned white coats, and in laboratories dismembered a plant until they had reduced it to its smallest working part—a single cell—and found that even this fragment of plant organisation could grow into a plant, complete with roots and shoots, leaves and flowers, before producing seeds from which more plants could be grown. In this way they discovered totipotency—the resonant and evocative name by which they recognised the ability of any cell to re-create a whole plant, provided it remains alive, and provided it is capable of division to rebuild new tissues and then new organs and finally a new plant.

Plants grow and develop by repeated sub-divisions of their cells, followed by expansion, growth and then—as each cell becomes specialised to play its own particular role—differentiation. The cells which initiate these events, by dividing, are not distributed throughout the entire plant but are restricted to particular positions, found repeatedly in all kinds of plants, from which all development and growth proceeds. They are extremely important to anyone propagating plants from cuttings of all kinds, because they are the centres of production of the cambial cells, the building bricks from which the new plants will be constructed.

8

Cambial cells divide into two, and having done so, one of the divisions retains its ability to divide, and, by subsequent divisions, enables the plant to produces new parts or replace old ones. The other increases in size and develops in a controlled and organised way. It becomes a cell of a particular kind with a well-defined function appropriate to the tissue in which it occurs, and in a form which is precisely characteristic of the plant of which it is a part. This is so commonplace that we take it for granted that a plant nearly always produces leaves in places where we look for leaves, of exactly the shape that we expect to find on the plant concerned. So, during the normal processes of differentiation that form the tissues and structures which compose a plant, the regenerative cells in the shoots divide to form stems, leaves, flowers, etc., and the corresponding cells in roots divide to produce yet more roots, and sometimes tubers. Yet every single cell within an individual plant shares exactly the same genetic constitution whether it forms a part of the leaves or roots, flowers or tubes, or whatever. It is its position in the plant which interacts with the cell's genetic constitution to decide the precise way in which it will develop.

Under normal conditions, most of the time, newly formed cells occupy a well-defined position, and pursue a predictable course of development, building up the tissues of roots or stems or flowers as the plant grows. This situation is altered immediately a cutting is made when a piece of shoot, or perhaps a section of root, is isolated from the rest of the plant. The act of separating it creates highly abnormal circumstances and the absence of vital parts, such as roots or leaves, which are needed for effective functioning and survival, can be recognised by the tissues within the cutting. As a result regenerating tissues located in appropriate parts of the cutting are reprogrammed to produce whatever new structures are needed to restore a complete and fully functional plant. This is the practical consequence of the phenomenon of totipotency, and it is this capacity for total renewal of an entire plant from whatever bits and pieces may be left, that the gardener relies on whenever he propagates plants by cuttings of any kind. But before anything can happen at all, the part of the plant used as a cutting must contain cells which are capable of division—most mature leaves contain no such cells, neither do tubers, nor flowers.

It may be encouraging or discouraging, depending upon the mood of the moment, to reflect that failures to propagate plants from cuttings are not due to the plant's incapacity (though some are much more capable than others), but to the gardener's inability, either to choose the right material, or to provide the conditions that enable a cutting to make its own repairs and renovations.

Adapting to the Environment

Plants either produce seeds, or reproduce vegetatively, to make up for losses as individuals die. Their success depends on their ability to interact favourably with the conditions in which they are growing. Put simply, tender plants would not persist in areas which experienced severe frosts; plants dependent on the constant availability of water—water lilies perhaps—could not grow in places where all the ponds dried up completely from time to time. At a slightly more complex level, though, there might be ways in which tender plants could adapt and survive in frosty places; perhaps by responding to the changing seasons by producing seeds before being killed, so that new plants would appear to replace their parents once the frosts were over. Or, water lilies might be able to survive periods of drought if they could perceive their onset, cease growth and become quiescent before the ponds in which they lived became so dry that actively growing plants died.

Survival in Nature

Plants respond continuously to their environment, and they do so whether they are growing actively or passing the time in an apparently passive state such as a seed or bulb or the almost buried crown of an herbaceous plant in winter. These responses may be direct and very visible. A leafy plant will wilt as soon as it is short of water, a seedling will grow rapidly in warm conditions, or develop unhealthily skinny and pale when light is inadequate. The responses may be indirect, and much less obvious. Internal changes in resting buds during the winter enable them to grow with extra vigour when warmer weather arrives in the spring. Herbaceous perennials produce strong anchor roots, or tubers, during the declining daylengths of late summer and early autumn, which they pack with starch to enable them to survive the coming winter and support vigorous fresh growth when the young shoots respond to the return of spring,. and their survival depends on being able to compete in a race for light and space amongst a crowded community of plants.

In the competitive state of affairs found in natural plant communities, plants need to be clairvoyant to survive: not only must they be able to make the best of what is going on around them, but they must also be able to interpret what is happening in ways which prepare them for future events. These subtle, often undisclosed, responses have to be taken into account when we propagate garden plants. It may be necessary to sow seeds a long time before they are expected to germinate so that they are able to make internal changes in preparation for the event. The best time to divide a plant may depend on the invisible activities of the roots much more than on what is going on amongst the leaves and flowers above ground. Hidden annual cycles of cell division in the shoots may define times when cuttings produce roots easily, and be the reason at other seasons why no amount of effort, or expenditure of time or money, will tempt a single root to appear.

Under natural conditions a species will be found in a particular spot only if it can endure the stresses and make use of the opportunities that occur there. These include the threats that anyone who gardens thinks of almost instinctively—climatic events like drought, frosts, searing winds or dank dampness. They also include the problems experienced by individual plants attempting to establish themselves and to persist in competition with the vegetation around them, and against the attacks of animal predators and fungal pathogens. Very many species can survive, day to day, in a place, and even from year to year for generations but, unless absolutely at home in their surroundings, may succumb to occasional abnormally intense periods of cold or drought, or the massed attacks of predators and pathogens in 'plague' years and lack the resilience to recover.

Survival in Gardens

Plants in gardens benefit from the sheltered accommodation provided by us to protect them from many of the stresses which they would have to endure in the wild. We try to eliminate competition from other plants, which is literally overwhelmingly important to the survival of wild plants, by spacing and careful weeding. The unwelcome attentions of predatory insects and infections from pathogens can be curtailed. Indeed, a very high proportion of the exotic plants now in gardens in temperate regions, arrived free from their natural pests and diseases, and do not provide a very attractive diet to native varieties. The fertility of the soil and its physical condition can be maintained or improved to encourage plants to grow vigorously, and the sharp edge of winter cold, or the withering effects of summer drought, can be relieved by shelter, by watering or by mulching. Most important of all, perhaps: when, from time to time, plants growing in gardens are destroyed, we can replace them with cuttings overwintered in a greenhouse, or by raising a batch of seedlings from seed gathered during a previous harvest.

These quite commonplace gardening activities are highly effective; just how effective becomes clear when the vast number of exotic plants which can be found in gardens is compared with the very small number which have escaped, and have naturalised themselves and been able to establish and spread as persistent wild populations in the countryside. There are notable examples of successful escapees, like the giant hogweed; the wild rhododendron; the Japanese knotweed; or the butterfly bush, which, instead of sheltering leopards in its home in central China, now provides a most attractive mantle to mask waste sites and ruined buildings. These conspicuous examples should not hide the fact that they represent a tiny proportion of successes; endowed with unusual ability to cope with the stresses which are the natural lot of the wild plants that survive in regions with a temperate climate.

Good gardening works because it reduces the stresses experienced by the plants being cared for, and as a result we can grow in our gardens plants which grow naturally in distant places with climates quite unlike our own. The more successful a gardener is in placing plants, and growing them in ways which suit them, the less stress they will suffer. In exactly the same way, propagating plants successfully depends on providing seedlings, cuttings or divisions with conditions which cause them the least possible stress. Unfavourable temperatures, inadequate illumination or excessive humidity may stress young seedlings and reduce their resistance to fungal attack. Fluctuations in humidity, high temperatures, dry atmospheres and soggy composts make life more difficult than it need be for cuttings and reduce their ability to regenerate new tissues or resist infection. Even seeds, those most seemingly inert forms of plants, suffer stress under unfavourable storage conditions, and quite easily lose their ability to produce seedlings.

Plants as Clairvoyants

The kinds of plants which make up the natural vegetation anywhere are there because they are the ones able to make the most efficient use of day-to-day conditions in that particular place, and, at the same time, prepare themselves for prospective opportunities or hazards. By far the greatest number of the plants in British gardens for instance, come from temperate parts of the world in which they experience very distinct seasonal variations in climate. They respond to the changing seasons by alternate phases of rapid growth and development while conditions are favourable, and periods of restricted growth or apparent inactivity to provide for a bare survival during seasons which bring with them severe cold, or extreme drought. The ways plants foretell the future to prepare themselves for what is to come are much less obvious. So much so, indeed, that the effects of daylength, the most significant of them all, was discovered barely 70 years ago.

Even though we are warm-blooded mammals, we still shiver when it is cold, and become uncomfortably hot in a heat-wave. Plants, with no system like ours to regulate their body temperatures, experience the ups and downs of changing temperatures a great deal more acutely. As we are all aware, temperatures change with the seasons, and when they are recorded and calculated as yearly averages they increase or decrease neatly and progressively from month to month. Plants are able to use these changes in temperature to assess what is going on around them and as an indication of future developments, and seeds, and the resting buds of trees and shrubs frequently depend on temperature to provide them with signals which alter the ways they germinate or grow. But temperatures are notoriously inconsistent from one day to another, and from one season to another, and erratic fluctuations and repeated occurrences of 'exceptionally' or 'unseasonally' cold or warm weather make temperature an inconsistent and unreliable indicator on which to depend too strictly. Rainfall, although in some areas extremely seasonal, is also very liable to unpredictably erratic variations.

Daylength

Rainfall and temperature are two highly noticeable features of the environment, and directly affect our comfort. We refer to them as the weather, and react to them in a very positive way. A third feature, which affects us just as fundamentally, but, for many of us less obtrusively in our daily lives, is the length of the period of daylight between sunrise and sunset. The seasonal variations in daylength are not only exceedingly marked in temperate parts of the world, but, what is much more important, they are consistent and predictable.

If on Midsummer Day we could find nothing better to do than rise as the sun rose, set our stop watches and wait patiently to discover that the sun disappeared beneath the horizon exactly 18 hours and 53 minutes later, anyone who knew about these things would be able to assure us that we were in Leningrad, or Oslo, possibly in Lerwick but, at any rate somewhere very close to 60° north of the Equator. At any place on the Earth's surface, the period between sunrise and sunset is the same on any particular date this year, next year, and every year afterwards, to all intents for evermore. The rates at which the daylength increases during certain seasons, and decreases during others, as well as its longest and shortest duration in mid-summer and mid-winter, are also precisely regular from one year to another. They will appear to vary a little depending on the presence or absence of thick cloud, but, taking one day with another, such effects are quite trivial. These seasonal variations in day-length change characteristically and regularly with latitude, and the polar lands, of the midnight sun in summer and perpetual darkness in winter, contrast with the Tropics where the days scarcely change in length throughout the year.

The consequence of all this is that if you or I take the trouble to record the daylength, the rate at which it changes from day to day, and its relative duration at different seasons, we can find out not only the latitude in which we are living, but also the season of the year, and consequently what is most likely to happen next. Plants can do this. They possess light-absorbing pigments in their cells which measure variations in the amount of light in specific narrowly defined wavebands of the spectrum, which provides them with a system which not only measures the length of each night, but also compares the length of successive nights. This ability provides the most significant way that plants growing in temperate parts of the world react to the changing seasons and prepare for the future, while making the most of the present.

The Subtle Effects of Daylength

The ways that plants grow and develop are very often controlled closely by changes in daylength. In response to decreasing daylengths embryonic flowers may be formed, deep in the innermost parts of the buds, far in advance of their appearance, but in time for them to emerge and flower, and produce seed at a favourable season. Similar responses enable annual plants to assess how long a season they have in which to flower and form seeds. Early in the year they will develop large leafy plants with many stems capable of carrying quantities of flowers; later as less favourable conditions become more imminent they will run precociously to flower and produce a few seeds at least before their time runs out. As daylengths decline in late summer and early autumn many perennial plants, trees and shrubs respond by setting in train the complex, co-ordinated changes which prepare them for winter. Growth slows down and ceases; storage organs are formed, and storage compounds, like starch and fats are diverted to them; complex organic molecules including chlorophyll and proteins are salvaged from the leaves, broken down into simpler compounds and moved back to be stored for the following year; the leaves become senescent and fall; resting buds are formed, packed with starch, and sealed within their bud scales to protect them through the winter. The constitution of their cells alters and their membranes become more permeable: changes which protect the cell contents from becoming frozen solid when winter temperatures fall, for long periods, far below freezing point.

Many of these responses are subtle and invisible; there is no outward change in a plant's appearance when the apical meristem starts to produce flower buds in place of leaves. But their effects on the plants are far-reaching and fundamental. Not surprisingly, it is vitally important when propagating plants to be aware of how strongly plants respond to changes in daylength, and what a dominant component of the environment these changes are. The late summer and early autumn are busy times for the propagator; sowing seeds of hardy perennials, taking cuttings of shrubs, dividing herbaceous plants. It is a time when seedlings, rooted cuttings and newly made divisions are being produced, and a time when, very soon, these young, barely established plants will be faced with the onset of winter and the problems of surviving until the spring.

Although still small, and apparently very vulnerable, they are able to respond naturally to the declining daylengths, and in most cases survive the winter in a resting state, to grow away strongly the following spring, making use of the slender reserves which they were able to produce and store the previous autumn. Attempts to force their growth, by keeping them in a heated greenhouse late into the autumn, or even right through the winter, are likely to be thwarted by the plants' sensitivity to the changes in daylength which mark the changing seasons, and are almost certain to do more harm than good.

The British Isles, though close to a continental landmass, are surrounded by water on the edge of a great ocean. The oceanic climate that results from this position tempers the heat of most summers, and the cold of most winters. It also produces variable and unpredictable conditions from one year to another. The treacherous nature of the climate poses problems to the plants which have to survive it under natural conditions, and the native British flora, which is not a very varied one, is restricted to those plants which can cope with its uncertainties. Paradoxically, the British Isles also owe the enormous variety of plants that can be grown in gardens to this same oceanic climate and its tendency to moderate extremes of heat and cold. Only occasionally are plants, growing in the favoured conditions we provide for them, subjected to the overwhelming shocks from extreme drought or intense cold, which would make their survival impossible. Consequently British gardens contain plants from exceptionally varied parts of the world. Many of them live naturally in areas where the climates are not just different, but fundamentally different, and these survive by accomplishing a compromise between their responses, programmed for other circumstances, and the prevailing conditions in which we grow them. The ways we look after our garden plants, and treat them during our attempts at propagation, will only be successful if these compromises are taken into account.

Plants: Widespread or Rare

Some of the plants cultivated in temperate areas can be found growing wild only in a very few places in the world, others are widespread and seem able to cope with all sorts of different climatic conditions and situations. Some are easy to grow, some we can keep alive, tenuously, with much difficulty and endless care. Plants which are rare in the wild, or difficult to grow in gardens, and the two do not necessarily go together, may be specialists. Some alpine plants, and especially those from high mountains within the Tropics, endure extraordinarily severe climates scarcely matched elsewhere, but do not thrive in what would seem to be much more benign lowland, temperate parts of the world. Other rare plants may be species with ineffective, limited capabilities; relics perhaps of populations which were once much more abundant or widely spread, but now hang on only in a very few places which happen to be favourable for them, and in which they are established. Widely distributed and easily grown plants have acquired the resilience typical of the true survivor, with a broad range of tolerance to a variety of conditions, based on genetic constitutions which interact flexibly with whatever is going on all around them.

It is not necessarily as easy as might be expected to propagate the plants which are widely distributed naturally, or which grow without difficulty in gardens.

Nor are rare plants necessarily hard to propagate. The ways that plants reproduce and the ease or difficulty of propagating them depend on the inborn nature of the plant involved, and this quite frequently produces surprising, even contradictory, situations. Some of our commonest, and most widely distributed, broad-leaved trees including the beech and native oak are practically impossible to propagate from cuttings of any kind, and we depend on seedlings as a means of raising young plants and on grafting for propagating specially selected forms. Japanese anemones, which spread like noble weeds in many gardens, usually transplant reluctantly when divided, and the best way to propagate them is from cuttings of their roots. On the other hand an extremely rare plant, *Silene viscariopsis*, whose world population appears to exist precariously on a single mountain in southern Yugoslavia, sets seed freely when cultivated, and its seedlings can be raised and plants grown to produce flowers with no difficulty at all. The dawn redwood, which has a natural distribution restricted to a handful of trees in western China, is one of the easiest of all conifers to propagate from seeds or cuttings, and grows very rapidly and obligingly in gardens.

The Origins of Garden Plants

All the plants in our gardens have been developed from wild plants, although some, like lettuces, tomatoes and onions, have been cultivated for so long that their wild ancestors can no longer be found. Some, like roses, daffodils, auriculas and gladioli, are hybrids with a complicated ancestry, derived from several different species, whose pedigrees provide enthusiasts with an endless subject for speculation, wrangling and debate. Very many others are attractive or pleasing species which we have adopted directly from their natural homes, and grow unchanged in our gardens. Whatever its origin, the parts of the world in which a plant grows naturally strongly affects its pattern of growth, the ways it responds to any climate when in cultivation, and, very often, the most satisfactory way to propagate it.

Garden Plants from around the Mediterranean

A disproportionately high number of the vegetables and flowers grown as annuals in Britain came originally from places not far from the Mediterranean Sea, or from parts of the world which have Mediterranean climates; and these include a small part of South Africa, Chile, California and Western Australia. These areas with well-defined climates challenge the plants which grow in them with conditions that are very unlike those found in the British Isles, and we might wonder why plants from such places should succeed so well in British gardens. Summers are hot with long periods of drought, which can be so intense that the land dries up as though it were a desert. Even the effects of occasional thunderstorms will be transient. Winters are cool, there may be occasional frosts but these are neither severe nor prolonged, and it rains, sometimes heavily. The great majority of plants, particularly the annuals, are incapable of surviving during the drought of summer, and have a winter growing season; they get through the summer as seeds lying on or just below the ground. These germinate in the autumn as temperatures fall and the rain soaks the ground again. Seedlings emerge and grow throughout the winter to produce their flowers late in the spring. By the time the drought of summer returns to shrivel up the plants they have completed their lives and their seed lies in the ground, waiting, to emerge the following autumn.

Thus, such annuals have evolved a straightforward strategy: their seeds germinate when temperatures are low, but will not risk emerging when the soil is warm; they germinate very rapidly; and extremely high proportions produce seedlings as soon as conditions are favourable.

These are very significant qualities for the needs of gardeners and farmers around the Mediterranean, and in almost any other part of the temperate world. Rapid germination at low temperatures of almost every seed exactly meets the specification for the seeds of a crop to be sown in the cool, even cold, soils of early spring in gardens and fields. These qualities were largely responsible for the success achieved by primitive arable farmers when they first deliberately tried to grow the grasses and other plants which eventually became major crop plants. They continued to be equally important as the practice of cultivating plants spread gradually to colder more northern places. And they retain their importance to this day. Even now, when we have at our disposal plants from parts of the world which were not even imagined in earlier times, plants originating from around the Mediterranean still hold their places in our gardens.

These regions of summer drought are also the homes of many of the aromatic herbs and shrubs we grow, particularly those with silvery or felted leaves, and some with broad, rigidly tough, evergreen leaves. These are plants well-adapted to survive heat and drought, and cuttings of silver-leaved plants, such as *Artemisia, Helichrysum, Santolina* and lavender, will all produce roots easily during late summer and early autumn, and continue to grow throughout the winter if given a very little protection from severe spells of cold weather. They are naturally adapted to survive dry conditions but are susceptible to wet, and succumb rapidly to disease if cosseted in the ways we are accustomed to use when looking after the softer shoots of deciduous shrubs taken as cuttings during the summer.

Survival in the Woods

The vegetation of the British Isles, left to itself, would return gradually but relentlessly to an enormous forest, dominated by deciduous trees, and blanketing practically all the lowland areas and all but the upper slopes of the mountains. Forests of this kind thrive in places where rain falls through the summer in sufficient quantities to provide for masses of foliage; broad green leaves have little resistance to desiccation, and soon stop functioning effectively during drought. Deciduous woodlands, most with a far richer variety of plants than can be found in Britain, occur in areas such as the continental parts of China, North America and Europe, where the winters can be very cold; in other parts of the world like the British Isles, Japan and the Pacific fringes of Canada and the USA, the influence of nearby oceans produces more temperate conditions.

Generally speaking the climate in these areas includes no devastating killing season like the Mediterranean summer, and a quick look at how green everything is suggests that conditions are much easier for plants. However, this impression is misleading. Individual plants are continuously involved in remorseless competition with their neighbours, which is extremely severe, and existence is a continuing struggle during which survival depends on the ways that different plants are able to make use of conditions around them. Seeds provide one way that plants can move about to seek out suitable spaces in which to germinate, within an environment which changes continuously as the dominant partners, the trees, grow up, decay and finally collapse. Consequently many species produce seeds with structures which enhance their mobility or which possess features, like berries and other fruits, which make them attractive to birds and mammals. These eat the fruits and later distribute the seeds far and wide. Many of these seeds possess rather precise means of controlling their germination, which enable them to take account of their location, and the time of year, so that they are more likely to produce seedlings in situations, or at seasons which favour their chances of establishing themselves.

Woodland Perennials

Many of the trees and shrubs and herbaceous perennials which can be found in gardens grow naturally amongst deciduous woodland. However, this woodland home, unlike the Mediterranean, provides no clearly defined 'best season' when seedlings can emerge. Seedlings, whenever they appear, are likely to encounter a balance of advantages and disadvantages depending on the severity of the winters, the activity of predators and pathogens, and the pattern of growth of neighbouring plants. In continental areas the climate tends to be fairly consistent, and certain seasons provide better chances of survival than others. Oceanic parts of the world, which include the British Isles, are notable for extremely inconsistent climates in which the prospects of survival vary enormously from one year to another.

The seeds of woodland plants mature from mid-summer onwards. The ground they land on may be moist and the weather warm, and seeds may either germinate immediately or remain as they are till a later date. Those which germinate at once have prospects of growing into small plants before the onset of winter and of being well-established the following spring. This gives them a chance to compete effectively with the neighbouring vegetation; to flower early in the season, and develop large plants capable of carrying numerous flowers. These are the advantages of immediate germination. The disadvantage is that they must face winter, and all its dangers, early in their lives when they are small and very vulnerable; their chances of survival, on the whole, are tiny. Those seeds that delay their germination until the spring pass the winter slightly more securely lying inconspicuously in the soil, but they may pay dearly for this when the young seedling finds itself struggling at a great disadvantage against well-established neighbours.

In some years the climate will incline the balance in favour of seedlings which emerge immediately, at other times seedlings which delay their germination until the spring will have better chances of survival. The prudent strategy, and one possessed by many plants growing in these areas, is to produce crops of seed which do not germinate all at once, but provide successive flushes of seedlings, or even a dribble of individual seedlings spread over many years. These more complex patterns are achieved in a variety of ways, and some of them seem to be designed to frustrate gardeners attempting to produce a nice even batch of seedlings.

A walk amongst woodland tells us at once what season of the year it is: the fresh foliage of young leaves in spring, the heavy, leadenly dull greens of summer, autumn colours, and then the bare, leafless branches that transform the scene through the winter. The changing seasons are a very dominant influence in regions where deciduous forests grow, and are reflected in equally definite cycles in the growth of the roots and shoots of shrubs and perennial plants. Cuttings taken from actively growing leafy shoots during the summer often produce roots rapidly, but they also suffer severely if they are neglected, particularly if they dry out. Their needs are quite unlike those of the silver-leaved Mediterranean shrubs, which thrive on the dry conditions to which they are adapted. These softer, broad-leaved cuttings only survive in an atmosphere which is constantly humid. Later, when their leaves fall, similar shoots, by then older, harder and leafless, can be used to produce hardwood cuttings which have no need for humid atmospheres and careful attention.

We cannot help noticing the appearance of leaves in spring, or their flaming departure in autumn, but almost every one of us passes by unaware of the ways that the roots, hidden beneath the soil, wax and wane with the seasons. These hidden cycles of root growth and development can influence attempts to propagate plants at least as decisively as the condition of the shoots. The activity of roots defines those times when plants can be divided successfully—particularly the ground flora—the herbaceous perennials of deciduous wood-lands. Quite a high proportion of these are practically evergreen. They take advantage of the shelter of the leafless trees, by remaining active through the winter months making a living during interludes of benign weather. Others renew growth precociously early in the spring. Some, like the cow parsley, can grow quite remarkable rapidly at low temperatures and manage almost to come into flower before the leaves start to unfurl on the shrubs and trees above them. During summer many of these ground-covering perennials become quiescent, shaded beneath the canopy of leaves. This phase of above-ground inactivity can be misleading. Below the ground the roots are starting their annual cycle of development, producing strong anchor roots during late summer and early autumn, which they use as storage organs for starch and other reserves during the winter. This is a time, almost universally unobserved, and so neglected by gardeners, when such plants can be divided, even though many will be fully foliaged, and will re-establish very quickly to produce strong well-developed crowns before winter comes.

Coping with Grasses

The Steppes of Eastern Europe and southern Russia, the prairies of North America and, in the southern hemisphere, Patagonia, are parts of the world from which garden plants have been collected, and which share a common feature. In all these places grasses dominate the natural vegetation; taking over wherever the rainfall is insufficient to provide for extensive stands of trees and shrubs. In these temperate grasslands the most characteristic features of the climate are cold, rather dry winters, usually with a covering of snow for much of the time, and hot summers during which periods of drought are broken at intervals by moderate falls of rain. Broad-leaved herbs growing amongst the grasses often form a conspicuous and beautiful feature of the landscape. Many have found their way into our gardens, many more would if they could endure the damp, muggy and dripping British winters, in which slugs thrive and steppeland herbs disappear without a trace. The prospects of growing these plants can be much improved by taking into account the ways they survive in the wild, and their natural seasons of growth and development.

Grasses compete formidably with any plant, even trees. Their slender leaves and thread-like roots convey an impression of fragility which belies their real natures. They crowd out and overwhelm the plants amongst them, and their dense filamentous roots are adept at removing water and nutrients from the soil, leaving their competitors deprived. Under suitable conditions they establish a dense sward which provides few opportunities for invasion by outsiders, particularly during the spring and early summer when the grasses are growing strongly. Later, after they have flowered and shed their seeds, many grasses stop growing for a while, particularly during periods of drought. The sward becomes ragged; gaps may develop and provide spaces, which give seedlings a chance to establish themselves when rain comes, provided they germinate rapidly and grow quickly. Most of these spaces will be colonised by seedlings from the abundant newly shed grass seed, but they also give broad-leaved herbs a chance. The success and survival of the herbs depends on how closely they synchronise their development with the grasses around them by producing seeds by mid-summer in time to occupy spare spaces when they occur. They are more likely to succeed if their seeds are able to germinate at high soil temperatures under rather dry conditions, as soon as they fall from their parents and compete aggressively with the grass seedlings in the race to establish entrenched young plants before the onset of winter. Seedlings persist through the winter sheltered from cold and winds beneath the snow, and when spring comes the young plants are well enough established to have a chance of surviving the competition from the surrounding grasses. They are able to grow away strongly using the reserves stored in their roots and crowns the previous autumn, and may even grow large enough to produce flowers and seeds during their first summer.

Many of these steppeland plants are not difficult to grow in gardens—just difficult to keep alive, and the main problem is their susceptibility to death and decay in the winter. In cultivation they respond well to being sown during late summer, to being allowed to grow into small plants crowded together in the containers in which they were sown, and left like that right through the winter. They are tolerant of rather dry conditions through the winter, and when the spring comes establish well if turned out of their containers, potted up individually and encouraged to grow on strongly through the early summer.

A Place in the Alps
Plants that grow naturally in mountainous parts of the world are called 'alpines' when grown in our gardens, irrespective of their true origin, and aficionados look on them as an elite group. Many are extremely attractive plants, which need no further 'improvement' by man to encourage us to grow them, and there is a tendency to look on hybrids a little disdainfully, as though inferior to the species, unless they happen to be forms that we have raised ourselves. The places they come from are too diverse to attempt comprehensive generalisations about their behaviour. Some, like those from high altitudes in the Tropics, are notoriously reluctant to conform as garden plants. They grow naturally in places where daylength scarcely varies from one season to another, and where

intense cold at night and high temperatures during the day expose them to a constantly repeated pattern which we cannot hope to counterfeit. Our climate, with its changes in daylength and strongly developed seasonal variations, is so alien to them that very, very few of these tropical alpines are able to make the adjustments needed to tolerate it.

The weather in almost all mountainous regions is harsh, uncertain and violent, and provides conditions where establishment and the occupation of a place on the mountain top become so important that they provide virtually the only hope of survival. Annual plants with their dependence on repeated re-establishment from seeds face a particularly hazardous existence, and are relatively uncommon. The perennials which grow in these places, whether shrubby or herbaceous, are more likely to survive by means of vegetative reproduction, during which young plants are able to derive support from their parents. Natural layers, the production of offsets, a tendency for stems to form roots easily are all commonly found amongst alpine plants and, while not totally replacing seeds, are likely to be more important as a means of survival than they are in less hazardous environments. Some plants rely on a form of propagation known as vivipary, more accurately but less evocatively known as proliferation, during which seeds develop into tiny plants before they disperse, and are lowered to the ground on the old flower stem, which continues to support them with water and nutrients while they establish themselves. Plants, growing in the ever-shifting heaps of stones that form screes, make positive use of their repeated burials by branching out and ramifying through the stones above them so that eventually the origin of the plant may lie many metres below the surface, and its branches survive as a more or less independent network of small plants emerging through the loose debris around them. As time goes by parts of the plant become separated but continue to grow independently, until large colonies develop, apparently of separate plants, but in reality all sharing a common source.

Many other alpine plants remain compact with densely packed, congested shoots, or overlapping leaves that protect the plants like tiles on a roof. They are adapted to tolerate exposure, to high winds and low temperatures—often the one reinforcing the effects of the other—but also, very often during summer, to high temperatures and intensely bright light as well. They are likely to be less well able to endure periods of high humidity and low light, combined with moderate to warm temperatures. These qualities and limitations very quickly become apparent during attempts to propagate alpines. Many rot away remorselessly in confinement, but can be propagated more easily unenclosed apart from the airy protection of a partially open cold frame or well-ventilated greenhouse.

Although seeds of alpines play a lesser part, they are still a significant means of reproduction, and are very important as a way of raising the plants we grow in gardens. Alpines may produce quite small crops of seed which will germinate only in response to particular conditions, interacting with the natural environment to ensure that seedlings appear at a favourable time. This is almost always the spring, and it has become normal practice deliberately to sow these seeds very early in the year so that they are exposed to frost for a period before the time when they are expected to produce seedlings.

African Marigolds and Mexican Maize

We have gathered the plants for our gardens from all over the world: plants, whose survival in the wild depended on the possession of particular qualities complementary to the hazards and opportunities around them. In the course of cultivation many of these plants have been changed literally beyond recognition. No plant that grows wild produces tightly sheathed and impenetrable cobs like those of maize, and their development from the small, loosely wrapped heads of their natural originals makes a long and complicated story. The size, diversity of shape and range of colour of cultivated dahlias and roses exceeds, beyond imagination, any variations to be found amongst their wild relatives. Our staple grain—wheat—is an intricately complex hybrid between three species from two different genera which looks nothing like any of its progenitors. These changes resulting from selection by man, and recently by deliberate, scientifically inspired breeding programmes, have altered the way plants look and the way they crop in a search for desirable qualities like fruitfulness, size, colour and conformation of flowers, plant vigour or form. Our eyes may be dazzled by a bed of African marigolds, almost unrecognisable as the descendants of a modest wild yellow daisy, and take it for granted that these cultivated plants are in every way different things from the wild representatives of their species. But, these changes, highly visible though they may be, are not necessarily accompanied by equivalent changes in a plant's invisible internal workings. The strategies which the plant developed to enable it to survive in the wild persist through generations in cultivation, often with little or no change, and can be identified and used constructively by gardeners who take the trouble to find out where and in what situation the plants that interest them grow naturally.

CHAPTER THREE

Economic Use of Equipment and Facilities

Growing plants from seeds, taking cuttings and persuading them to produce roots, using division as a simple means of plant multiplication are fascinating and, some would say, amongst the most satisfying of all aspects of gardening. Part of this enjoyment can come from a feeling of getting something for nothing, which warms even the least worldly amongst us with a glow of smug satisfaction. This is a deceptive feeling to indulge, and must be responsible for many of the disappointments which gardeners have to go through when they first make attempts to grow their own plants. A kind of beastly morality looms behind the natural history of living things which ensures that nothing comes for nothing, and requires that a price of some kind must be paid for everything! The price may be paid with the care that plants need if they are to grow well; it may be levied by taking the trouble to learn about the plants, and discover more about the ways they should be managed, and the times of year when attempts at propagation are likely to be successful; it may be rendered by moderation in balancing ambition against the means available; and means is not very often synonymous with wealth, but much more likely to have a lot to do with the skill, involvement and experience of the propagator.

Economic Use of Facilities

The feeling of getting something for nothing does not always come from failing to take time and care into account, but can be due to a blind refusal to look at very real financial costs. A familiar example is the half dozen geraniums, dug up before the winter and overwintered in a greenhouse which has to be kept warm to prevent frosts from destroying the plants. Greenhouses do cost money, and they deteriorate so that one day they need to be replaced. The cost of even minimal heating to keep them frost-free can be very high. The space inside them can be used much more effectively in other ways. Adding one thing to another it is not unusual for the real costs of overwintering a few geraniums to tot up to £10 or £20 a plant, quite apart from a good deal of time and effort. This would be more than enough to finance a spending spree at the local market in the spring to fill the garden with a mixture of annuals, geraniums, fuchsias and petunias; sufficient to keep the garden in colour throughout the summer. The presence in the greenhouse of a few other forlorn plants which are hardy and would survive perfectly well elsewhere, piles of discarded seed trays, miscellaneous flower pots and plastic watering cans, together with bottles of insecticide and canisters of powder intended to destroy moulds and mildews—

none of which should be in a greenhouse, and do no good by being there—does nothing to offset this cost.

Greenhouses can be economic power-houses in a garden, and can be used exceedingly effectively to bring down the cost of gardening, and to produce plants very, very cheaply. The key lies in finding ways to use them, and to combine them with their indispensable ancillaries including cold frames and nursery beds, in ways which not only work, but which match the scale and needs of the user with very little waste. More often than not greenhouses are not used to the full because their potential is not understood and they are left half empty for months on end, particularly during the winter when they should be working at their hardest. An almost equally frequent source of wasted space and effort can be found in the containers used to grow plants, produce rooted cuttings or germinate seeds.

Salvaging Domestic Waste

Amateur gardeners are advised, time after time, to salvage redundant food containers, plastic cups and domestic rubbish of all kinds and to use this jetsam from their garbage bins as flower pots or seed trays. Often this advice comes from professional gardeners who would never consider using the things themselves. These suggestions are almost always misguided and, at best, are a way to lose pounds while saving pennies. The expanded polystyrene cups used to serve hot drinks look like perfect substitutes for small flower pots. The aluminium trays from the local take-away could suggest themselves for use as seed trays; any enthusiastic yoghurt eater could empty enough cartons in a year to provide pots for geraniums sufficient to overflow the garden. There are snags. None of these containers has any drainage holes, and these must be made. It is not sufficient to push a hole through the bottom and hope the water will find its way out—just how insufficient this is will be revealed by looking at the patterns of drainage holes in modern plastic flower pots. There will be a number of holes or slots in each, not only in the bottom of the pot, but also towards the bases of the sides. This array of holes is there because the quickest and easiest way to kill most plants is to grow them in a soggy, saturated compost. Peat-based composts, in particular, cause trouble unless the containers used to hold them are very well-provided with ways for excess water to drain away.

The sizes and dimensions of containers salvaged from domestic refuse are almost certain to vary. Like their owners, some will be tall and thin, others short and fat, and many more something in between. Some will hold a pint, others half a litre. Different volumes of compost in pots of different shapes dry out at different rates, and a batch of seedlings set up in a motley collection of salvaged containers becomes impossible to look after properly. Of all the arts and skills of gardening, one of the hardest to get right is the 'simple' task of watering, and it is much easier to do this well when the plants at least start off under uniform conditions; and putting seedlings into all shapes, sizes and conditions of different containers just makes things more difficult than they need be. Ill-judged applications of water mean that some plants will be too wet, others too dry, and before long these will cease to thrive and some will die. A very few plants lost soon outweighs all the small savings made by not buying well-designed, suitable containers and, during the long dark, dull days of winter, when watering becomes particularly difficult, whole batches of plants can be decimated if they are kept a little too wet for their liking.

Choosing the Container

Once thoughts of using domestic salvage have been abandoned, the problem remains of deciding which of the many containers available are the best buy, bearing in mind that they must be economical, practical and serviceable. Their prices vary widely, ranging from the very cheap, black polythene bags in which plants have been marketed for some years to relatively expensive pots made out of flexible polypropylene. Both can be used over and over again, provided a little care is taken when handling them, and can be stored anywhere that is convenient. Other forms of plastic, but especially polystyrene, are extremely brittle; they disintegrate suddenly, noisily and to the accompaniment of unseemly oaths, when they are picked up, and the plants they contain then fall to the ground and suffer dreadful damage. Their low cost does little to compensate for the frustrations they cause, and may not look economical at all when the short life of these containers is set off against the number of times that their more expensive rivals can be used.

Pots of Peat or Plastic

Products made from peat or paper-like materials have been marketed for a number of years as containers in which to grow or propagate plants. Their big advantage, so the salesman tries to persuade us, is that plants growing in them benefit because they do not suffer the disturbance of being turned out of their containers when they are potted on, or the time comes to set them out in the garden; they can simply be dabbed in, plant/pot and all together. These containers can be used only once, and this makes them relatively expensive, but, apart from that, the argument is less than convincing. There is no reason why plants should suffer when they are removed from their pots if this is done at the right time and without being very clumsy: even if it seems a little fraught to a novice gardener making tentative attempts with fears in mind of smashed plants and nothing left but a handful of disintegrating roots. Once tried it turns out to be a very simple thing to do. Plants treated in this way enjoy the great advantage, that, as soon as they are planted, their roots are in direct and immediate contact with the soil around them. The presence of a peat or paper pot can even reduce the chances of successful establishment, especially if the weather or soil is a little dry, when it is all too likely to act as an isolating barrier between the plant inside it and the soil beyond.

Another disadvantage of peat pots and blocks is that watering can be exceedingly difficult to judge accurately, particularly when a decision has to be made first thing in the morning before a daily absence at work. It is hard to avoid overwatering in winter and underwatering during periods of hot weather, and

---

Figure 3.1   Seedlings, cuttings and plants can be grown in many different containers; but propagating plants successfully, and economically, depends on matching the container used to the individual needs of different situations. A great deal of space can be wasted by using unnecessarily large containers, especially seed trays, and surprisingly large savings can be made by simple changes like using square pots in place of round ones.

Even a half-tray can hold a thousand seedlings or a hundred heather cuttings

Salvaged 'seed trays' such as egg boxes or take-away dishes are unsatisfactory substitutes

Plastic pots are easier to manage than traditional ones made of terracotta; square pots occupy less bench space than round ones.

Domestic salvage is a poor substitute for well designed pots

Polythene bags are cheap and effective, but take more time to fill with plants

Dry compost

Gap between peat pot and
dried-out compost

Protruding rim of peat pot acts
as a wick

Gap between peat and soil

Peat and compost dry out

New roots growing straight
into surrounding soil

Root-ball in direct contact with
surrounding soil

the effects of quite small errors in either direction gradually build up into problems. Once peat dries out, even if it contains a wetting agent, it can only be persuaded to take up water again effectively by thoroughly soaking it. The best way to do this is to set the pots out in wooden trays and float them like rafts in a tankful of water, until the peat becomes thoroughly saturated; merely watering heavily and repeatedly is mostly ineffective. Plants in peat pots or blocks on the edges of a group of plants, or in small isolated groups, are especially vulnerable and almost always show signs of the effects of drought. This can only be avoided by very careful watering, and it is a simple matter of technique always to start watering a block of plants by attending to the plants on the edges before doing anything about the centre.

It is inconvenient, or impossible, for most amateur gardeners to keep a constant, watchful eye over their plants and most are unwilling to buy, and should not need, the sophisticated semi-automated systems that commercial growers use to ensure the health and well-being of their stock. Containers which behave predictably, and which are simple to set up and use, are a boon, and make success much easier. Until the mid-twentieth century terracotta pots were almost the only ones available and there was virtually no choice in the matter. Today, we enjoy the benefits of living in the age of plastics, and have a good deal of choice. Many people retain an affection for the old-style clay pot and appreciate its feel and the way it looks, but, whatever the aesthetic arguments and however evocative of hand-skilled crafts they may be, there is no doubt that for most purposes they are less practical and more likely to cause problems when propagating or growing young plants than are containers made from plastics. Clay pots are heavy and clumsy to use; they stack unevenly; and have a gift for falling over and breaking themselves, even when they are stored carefully in dry, frost-free places. They are more likely to produce drainage problems, and the compost within them dries out more rapidly and more unevenly, so that they require careful management when used with peat-based composts. And they are expensive.

Almost the cheapest of all plant containers are black polythene bags with drainage holes punched through their bottoms, their most common shortcoming being the inexplicable reluctance of some manufacturers to punch out sufficiently large holes. Provided they are treated with a little care these bags can be used over and over again and occupy very little storage space. They appear to compare in cost extremely economically with the more expensive rigid containers, apart from one thing. The soft, floppy bags are fiddly to open and awkward to handle and to fill with compost. It takes longer to install a plant in one than in a more rigid container, and this difference becomes very significant in action in the potting shed. At least 100 rooted cuttings can be set up in rigid pots in the time it takes to put 70 into polythene bags. Few of us have unlimited time to spend on gardening, and the extra plants potted up in the same time using rigid pots can be the difference between getting a job done, and never getting around to finishing it at all. Any saving made in costs by using the cheaper container is likely to be outweighed many times by the value of the

Figure 3.2   Plants in peat pots can suffer from drought unless carefully handled when they are planted in the garden.

plants lost, and it can easily come to the point where £2 a plant is being lost to save 10p a pot!

Container Economics

Gardeners have been using clay pots for so long that they are still accustomed to conjure up the image of a round, terracotta-coloured, truncated cone whenever they think of a container for their plants. And the reason for round pots? Simply because, throughout the ages, almost all pots were round because of the way they were made. Plastic pots, manufactured by an entirely different process, can be fashioned in any shape which is convenient, and a number are now being produced which are square in section rather than circular. These occupy space more economically than round ones and are a more practical shape. A little time with a calculator shows that a circular container with a diameter of 10 cm (4 in) contains the same volume of compost as a square one of equal depth, with sides of $8\frac{1}{2}$ cm (3 in). The difference may not sound very large but it means that a space on the greenhouse bench which holds 200 plants in circular pots harbours 76 more, in the same volume of potting compost, in square ones. A change in the shape of pots being used can increase the space available by about a third. It would usually be much more economical to change the shape of the containers in which plants are being grown than to buy a bigger greenhouse!

Saving Space on Sizes

One of the commonest and easiest ways to waste space is to use containers which are unnecessarily large, and the relatively minor example just referred to showed one way to avoid doing this. But, literally the biggest offender is the seed tray—used automatically and thoughtlessly, even on television programmes intended to instruct, as *THE* thing in which to sow seeds. The standard seed tray, now always made of plastic, measures 851 sq cm (138 sq in), which may not convey a great deal to many of us. It conveys much more when measured by the numbers of plants that a seed tray can hold. It provides adequate, if confined, quarters for anything from 500 to 2,000 bedding plant seedlings, from the time they germinate until they should be pricked out. If these are the sort of numbers that are needed, then a seed tray is needed, but most of us need nothing like those numbers of petunias, pansies or penstemons, and even a half tray would be extravagantly large. Much more economical containers in which to sow seeds are small, square 7 cm ($2\frac{3}{4}$ in) plastic pots; these occupy a fifteenth of the room but each provides space for around 50 to 200 seedlings, which are enough for most purposes. Similarly, small plastic containers are almost always more economical than seed trays, and more appropriate to the quantities needed, when used to hold cuttings. Square pots with sides measuring 9 cm ($3\frac{1}{2}$ in) hold, as examples, about 25 heather cuttings, 16 lavenders and nine forsythias. These are the kind of quantities which would fit the needs of many amateur gardeners and seed trays holding something like 200 heathers, 100 lavenders and more than 70 forsythias are wastefully out of scale, especially if they are being treated to expensive heated space in a greenhouse or propagating frame.

The bedding plants that we buy are often pricked out in strips, or in cells, used to sub-divide seed trays, and this may seem to be a sensible idea, and one

Relative sizes/capacities of a variety of containers

Description of container	No. filled ex. 20 l. bag	Cost of compost*	Space (sq. m.) per 25 units	Bench capacity
7 cm square plastic pot	90	4.4	0.12	750
8 cm square plastic pot	55	7.3	0.16	562
9 cm square plastic pot	42	9.5	0.20	450
11.5 cm square plastic pot	19	21.1	0.33	273
9 cm diam. round plastic pot	58	6.9	0.20	450
13 cm diam. round plastic pot	20	20.0	0.43	209
17.5 cm diam. round clay pan	13	30.8	0.77	117
standard half-tray	11	36.3	0.96	94
standard seed tray	5	80.0	2.17	41

* This calculates the cost of compost in each container assuming a basic cost for a 20 litre bag of 400 monetary units.

'Bench Capacity' indicates the number of containers which would fit onto the benches of a 3 m × 2 m greenhouse, assuming that these occupied 60 per cent of the total area of the house, i.e. 3.60 sq. m.

Note that a square container with sides 11.5 cm holds very nearly the same amount of compost as a round one 13 cm in diameter—both hold about one litre. But 273 square pots fit onto the bench space occupied by only 209 round pots; in other words changing from a round to a square container increases bench capacity by 64 pots or 30.6 per cent.

Effects of container size
(comparisons within the range of square pots, using 7 cm as a standard)

Size of container	Reductions in numbers filled	Increases in cost of compost	Reductions in bench capacity
8 cm square plastic	39%	70%	25%
9 cm square plastic	53%	116%	40%
11.5 cm square plastic	79%	380%	64%

Figure 3.3. Seeds can be germinated and cuttings struck in a wide variety of different containers. Traditionally, seed trays are used but large economies can be gained by using smaller containers. The first table shows how many containers of various different sizes and shapes can be filled from a 20 l bag of compost, and the effects this has on costs and the numbers of containers that can be fitted onto a greenhouse bench. The second table shows effects of container size (using 7 cm square pots as a standard).

that we could use ourselves. It is, but it's not! The sub-divisions of the tray or the provision of a separate container for each plant enables the seller to market them in sections or small groups without those that are left behind suffering the damage that results when one or two plants are prised out of an entire tray. But, seed trays are shallow, and, when sub-divided, each cell contains very little compost; the cells do not connect one with another, and if, as a result of less than perfect watering, some are overwatered and others are merely splashed, the water in one cell cannot diffuse into another to balance out the supply. This system operates on very narrow margins between success and failure, depending on good management and good equipment to work. In a nursery where there is usually someone around to make sure all is going well it can give good results, but at home, where it may be difficult to keep an eye on the weather, and on the plants and on the watering all the time, the results can be disappointing. It is safer, and in most situations preferable, to sow seeds in small pots, which provide individual containers with a reasonable quantity of compost in each, and later to prick out into whole or half trays, depending on the number of seedlings being grown.

Propagating space, whether in a greenhouse, a cold frame, a heated propagator or bench fitted with soil-warming cables (even the kitchen window sill) is, like money, always in short supply and hard to find, especially at busy seasons of the year. Far the best way to avoid needless expense, and to make the most of what is available, is to think very carefully about the ways that space can be saved by using containers that match requirements but no more.

Space can be wasted equally prodigally when plants, which have been propagated, particularly shrubs or conifers, are being grown on to plant out in the garden or give away. A visit to a garden centre, where shrubs are sold in 3 litre (5 pint), or even larger, containers, can leave the impression that this is the size appropriate to their needs. In fact when rooted cuttings are potted up for the first time they should be fitted into the smallest containers that will comfortably hold them. In other words their roots should ramify through much of the space in the container when they are suspended in it immediately before the compost is added. This is not just a matter of saving pennies, but good gardening. Plants are very easily killed at this stage if they are over-potted and, it also economises on space especially when plants are to be overwintered in containers, when every square centimetre in a greenhouse or frame becomes precious.

Economising by Planning

Savings may be possible by planning ahead to take advantage of naturally warm seasons of the year, or to avoid having to use artificial heat to keep tender plants frost-free through the winter. There are also more opportunities than many people imagine to use methods of propagation which do not need expensive equipment like greenhouses, or mist propagation, or the continuous care and attention that elaborate methods demand. Amateur gardeners may often prefer to use methods which are dependable without great expense or effort, even though they take a little longer to produce a plant than some of the high input techniques now employed by professionals.

Until quite recently, for example, a sheltered nursery bed was almost universally the commercial practice for most shrubs. Today, it's almost as though it has been forgotten as an option, or is looked on as being second rate. For most amateur gardeners it provides by far the more economical, safe and effective alternative, when establishing young shrubs or seedling perennials, to potting them up into larger containers.

Setting up a Nursery Bed

A nursery bed will cost a little bit to set up, but almost all the materials needed can be second hand, or even well-chosen salvage, and costs and trouble will be more than offset once it is being used. The first significant saving is not having to buy compost to fill the large containers into which plants would have been potted on, to complete the later stages of their development. An even more welcome benefit may be the reduction in the amount of attention and management needed to grow plants successfully. Plants can be looked after with much less effort when they are more or less self-supporting in a nursery bed, than when they are growing in containers and depend on daily attention to keep them adequately watered and fed through at least one, and often two, growing seasons.

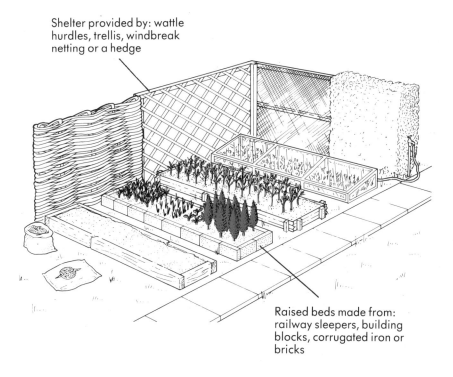

Shelter provided by: wattle hurdles, trellis, windbreak netting or a hedge

Raised beds made from: railway sleepers, building blocks, corrugated iron or bricks

Figure 3.4   Nursery beds set up in a corner of the garden are a practical and economical way to grow young plants. The beds should be sheltered in some way from cold winds, and are easier to manage if they are narrow enough to stretch across, and raised a little above ground level. Each bed can be filled with a prepared mixture of grit and loam, or top-dressed as freely as possible with grit, peat or spent compost.

Potentially, plants grown in containers, particularly if they are under cover, are capable of growing much more rapidly than those planted in the ground outdoors. But 'potentially' is an alluring word that is designed to raise great hopes and makes no promises, and in this case the potential is not easily achieved. Occasional periods of neglectful watering, or failure to feed the plants when they need it, very soon result in stunted misshapen starvelings—a disappointment rather than a fulfilment of hopes.

Thinking About Annuals

An early start with the sowing of annuals might seem to be the mark of a virtuous gardener. It almost always involves needless expense, and, unless very good greenhouse and cold frame space is available and time can be found to care for the plants skilfully during periods of poor light and cold weather, it can be very unproductive. Early April is soon enough to sow most of the annuals which provide our gardens with colour during the summer and, extraordinary though this may seem to some, early May is not too late at all. It is not necessary, and in fact quite wrong, to aim to grow the large, crowded plants which display themselves in flower in their seed trays in every garden centre, and on pavements outside ironmongers and newsagents in high streets. These are grown to show themselves off because that is the only way that annuals sell to customers who know no better, and who never learn, even though every article ever written about bedding plants condemns them. The plants will do much better set out into ground that has been warmed by the sun, at an earlier stage in their development, when they are still small and compact, and just waiting to grow away vigorously. Plants raised from seeds sown quite late in the season can be grown right through to flowering without the checks from cold weather which earlier sowings suffer, and without obligatory attendance on the weather forecast every evening, so that in a late frost all can be cosily enclosed in newspapers or sheets of bubble polythene.

There is much to be saved and everything to be said for thinking more about the annuals to be grown. Only a fraction of those available as seeds in the catalogues of seed merchants will be found amongst the ready-made bedding plants offered for sale in trays and sections every spring. The latter form a small coterie which share an assured popular market, and a great many of them, coming from sub-tropical and tropical parts of the world, can be grown successfully only at high temperatures; amongst their number are begonias, busy lizzies, dahlias, ageratum, petunias and scarlet salvias. One of the greatest advantages of growing annuals for oneself is that it opens the door to a more imaginative choice and the opportunity to include those that are less frequently seen. Many of these are easier to grow, and thrive at much cooler temperatures, than the stereotypes that appear in every other garden. The most economical annuals to grow are the ones which comes from places with climates similar to areas around the Mediterranean Sea (p. 46); almost all produce seedlings and develop well at low temperatures, and most survive a degree or two of frost without suffering catastrophically.

Shrubs from Cuttings

A great variety of shrubs can be propagated in the summer from cuttings taken between late June and the end of August, and this is one of the easiest, most economical, ways to propagate them; so much so, that anyone with only moderate ambitions who just wants to propagate a selection of shrubs without much trouble to fill up the garden, might well restrict all attempts at cuttings to this season. Commercial producers, more often than not, hustle these cuttings along a bit; encouraging them to produce roots by using soil-heating cables and by using mist propagation equipment to keep the leaves constantly moist. These methods prolong the season when cuttings can be taken, and speed up the process, besides increasing the chances of a successful result with some of the more demanding plants. But a great many shrubs can be propagated very effectively with no need for costly equipment of this kind provided cuttings are taken while outdoor temperatures are still fairly high, and they are set up in closed containers, which provide a water-saturated atmosphere around the cuttings as a substitute for mist propagation. A similar method, using sheets of plastic film to enclose cuttings, is now becoming so popular as a means of propagation in nurseries that in many places it is being used in preference to mist; this is also a very simple system to set up on a small scale.

Cuttings treated like this usually produce roots more slowly than they would with mist propagation and some help from soil-heating cables, and will probably not be ready to pot up until the following spring. Nevertheless, there is much to be said for it as a very economical way to produce plants; although it may seem a disadvantage that the cuttings are not ready to pot up until the following spring, in reality it is an added economy. Twenty cuttings kept together through the winter in the container in which they produced roots occupy barely a tenth of the space they would need after being potted up separately. It is far easier to find a sheltered place for one potful, perhaps in the front porch, or the window sill of an unheated bedroom, than to shift 20 individual small plants to somewhere safe when cold weather is forecast.

Equipment

Specialised equipment or methods have to be used to propagate some kinds of plants, and these will be described later. There are, however, more general ways in which equipment and facilities can be used economically in many different situations. Greenhouses are expensive items, which are often used very ineffectively. It is almost impossible to employ them to propagate or grow on hardy plants unless at least part is provided with enough supplementary heating to keep the frost out. The air space within the greenhouse can be heated artificially, using electric fan-heaters, a solid fuel or gas-heated boiler with heating pipes, or paraffin stoves. Whatever source of heat is used, space heating of this kind is indiscriminate and bound to be expensive. It is also not necessary: unlike us, plants do not move about, and spending money keeping the entire space within a greenhouse warm is not particularly effective when encouraging cuttings to produce roots, raising seedlings or overwintering young cuttings or seedlings. Much better results can be obtained, much less extravagantly, by using soil-heating cables buried immediately beneath the containers holding the cuttings or seedlings, or, on a very small scale, by using electrically heated propagators.

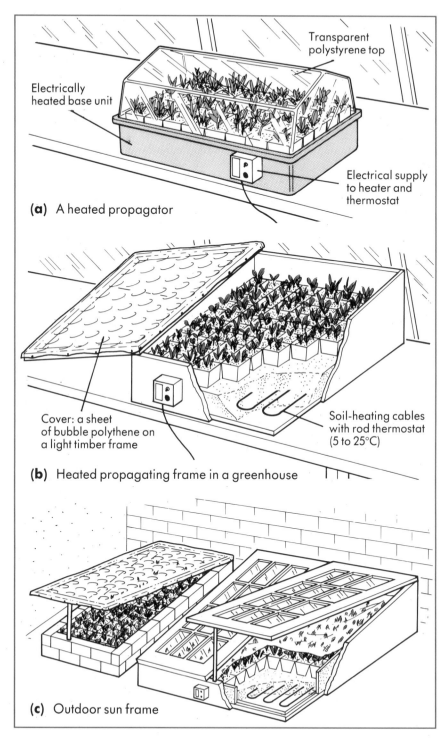

Transparent
polystyrene top

Electrically
heated base unit

Electrical supply
to heater and
thermostat

(a) A heated propagator

Cover: a sheet
of bubble polythene on
a light timber frame

Soil-heating cables
with rod thermostat
(5 to 25°C)

(b) Heated propagating frame in a greenhouse

(c) Outdoor sun frame

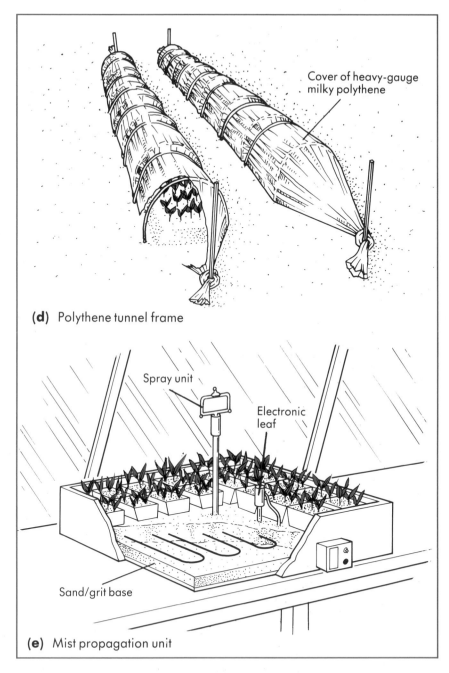

(d) Polythene tunnel frame

(e) Mist propagation unit

Figure 3.5 Cuttings can be kept alive in a variety of ways while they produce roots. The equipment used should be chosen to match needs, and the facilities available. Heating cables will greatly reduce the time taken by cuttings to produce roots. All electrical equipment must be professionally installed.

Using Soil-heating Cables

Soil-heating cables can be bought in various sizes, and can be set up on any scale, with thermostats installed to control their performance, to heat some or all the benches in a greenhouse; or they can be used effectively and economically to keep a well-insulated frame frost-free. Ready-made propagating units, most of which will hold the equivalent of about three standard seed trays, are usually conspicuously visible in displays at garden centres, and these suggest a tempting solution to the problem of providing heated space for seedlings or persuading cuttings to produce roots. A purpose-made installation constructed in a convenient place can provide so much more heated space for a comparatively small extra cost that it is almost always a much more economical solution than a ready-made unit, and is preferable for all but the most modest needs.

An area of bench, and it need be only one metre square (1 sq yd), fitted with soil-heating cables in an otherwise unheated greenhouse, is an exceedingly effective and versatile unit for propagating and overwintering plants. It can be employed during January and February to germinate seeds of hardy, herbaceous perennials, or to propagate cuttings of broad-leaved evergreens; between April and May for germinating seeds of hardy and half-hardy bedding plants and other annuals; between May and September as an aid towards

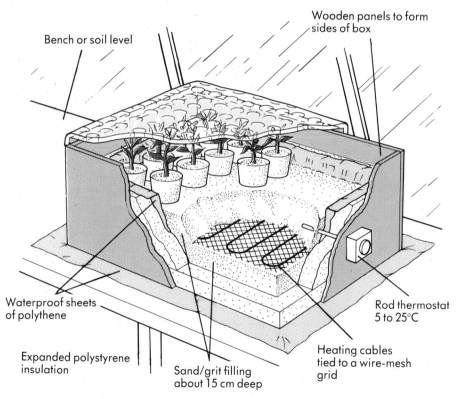

Figure 3.6   Thermostatically controlled soil-heating cables are the most economical way to provide warmth for cuttings and seeds.

encouraging rapid root development of a wide variety of deciduous shrubs; and finally, not only to encourage cuttings of silver-leaved and semi-tender shrubs to produce roots during the autumn, but also to improve their prospects of surviving the winter and being in good condition to plant out in the garden the following spring.

Insulation and Shelter

The most effective way to reduce heating costs and protect cuttings and seedlings from frost is to use sheets of bubble polythene, laid directly over the plants at night, and during periods of cold weather. This sandwiches the plants in a semi-closed container of warmth, which involves heating no more space than is absolutely necessary, and is much more effective, flexible and economical than attaching the sheets to the interior of the greenhouse. Bubble polythene can also be used to insulate plants in cold frames, not only as a blanket between the plants and the framelight, but also as additional layers of insulation pinned to light timber skeletons laid over the top of the frame. A cold frame provided with soil-heating cables and insulated in this way, with a thermostat to control heat input, provides a simple, unobtrusive way to overwinter plants of many kinds safely, even including the expensive half dozen geraniums that first stimulated this search for more economical ways to propagate and grow plants.

Plastic tunnels are a common sight on nurseries and are used to propagate and, especially, to bring on young stock. They provide invaluable shelter in which to grow plants rapidly during spring and summer, and to shelter them from cold winds during the winter, and are an economical solution to many problems when relatively large numbers of plants have to be produced. They should not be thought of as alternatives for a small greenhouse; though providing more space for less money, they are much more limited in the ways they can be used. Poor ventilation during the summer, susceptibility to cold and high humidity during the winter and the need for repeated renewal of the plastic skin all reduce the value of these tunnels, and detract from their effectiveness compared to well-ventilated and well-sited greenhouses. A small greenhouse can be very productive (see Chapter Thirteen) when used for plant propagation, and its versatility and effectiveness far outweigh the lower cost per unit area, which make polythene-skinned structures seem to be a bargain.

Matching Equipment to Needs

Any kind of equipment is only worth having when it is carefully chosen to be in scale with the use intended, and when it is backed up by other facilities which allow it to be used to its full potential. A ready-made electric propagator provides an example of equipment on which it is tempting to spend money in circumstances where it can scarcely be used effectively at all. Heated propagation space is a useful and worthwhile investment, for bringing on seedlings and encouraging cuttings to form roots rapidly, but only if it is sited in a well-lit and protected environment, such as a greenhouse or conservatory—not a window sill—where there is plenty of room to space out the cuttings or seedlings when they outgrow the propagator, and need to be potted up in larger containers. A greenhouse itself is also ineffective unless cold frames are available to protect plants removed from it, and give them time to acclimatise to colder, more exposed conditions before they are finally moved out to take their chance in the open air. The equipment needed to produce and grow plants forms a pyramid, based on unelaborate, simple methods at the base, and ascending through more complex and, inevitably, more expensive, facilities such as cold frames, greenhouses, soil heating and mist propagation, until it reaches at the apex the advanced laboratory resources needed for modern methods of meristem culture.

When the supporting facilities at the base of the pyramid are not available it is no better than a waste of money to buy the more elaborate equipment in its upper levels. However, a great many plants can be propagated very successfully using the simplest possible facilities, with no artificial heat of any kind, and with only minimum protection from the elements. Shrubs can be layered; hardwood cuttings taken during the winter will produce roots in sheltered corners of the garden; the seeds of hardy annuals, perennials, trees and shrubs can be sown outdoors in seed beds or containers. During the summer many cuttings will form roots very satisfactorily with no more protection than a sheet of polythene stretched on hoops above them. Perennial plants can be increased by division. These methods are all exceedingly well-suited to amateur use, both in their scale and their economy. They are methods that are sometimes disregarded—perhaps they seem to be too simple to be called propagation—in favour of more elaborate and expensive systems.

This is certainly not the case. The real art in propagation, as in other aspects of gardening, is to find a successful method which exactly suits the needs of the moment, without using more effort or buying any more elaborate or expensive equipment than is absolutely necessary.

CHAPTER FOUR

Directory of Propagation Techniques

An index is the conventional guide to the contents of a book, and is useful for those who know precisely what they are looking for. An index is less helpful, however, to anyone who is not so knowledgeable about a subject.

This chapter therefore provides a concise summary of the various ways by which plants can be propagated, and the conditions they respond to.

Chapters 5–11 each surveys the methods by which particular groups of plants may be propagated. The groups into which garden plants have been divided are detailed below, with the relevant chapter number shown in brackets.

Annuals (5): Conifers and Heaths (6): Alpines (7): Hardy Perennials, Grasses and Ferns (8): Shrubs (9): Trees (10): Bulbous Plants (11): Seed Collection and Storage (12).

No precisely defined divisions separate all these groups and gardeners do not always agree on the ways they use plants in their gardens. Ambiguous situations occur where, for example, plants included here in the chapter on alpines might be sought by others amongst the hardy perennials. Only a small minority of the plants covered are likely to give rise to problems of this nature, and, when difficulties are encountered, reference to the index should resolve them rapidly. The term 'Bulbous Plants', as used here, is not restricted solely to plants which produce bulbs or similar structures such as rhizomes, corms and tubers, but includes all those plants sometimes referred to as the petaloid monocotyledons. Again, any problems of interpretation which this might cause should be resolved through the index.

Throughout this book three major methods of propagation are described, and the ways that they can be used to propagate plants are set out in the order shown below in each chapter. The only partial exception being the annuals, in which methods of propagation are mostly confined to seeds.

Seed Germination.
Propagation from *Cuttings*.
Multiplication by *Division*.

Page cross-references, below and throughout the rest of the book, direct the reader to more complete descriptions, or additional information about relevant methods of propagation.

Section One: Seed Germination

Attempts to germinate seeds successfully depend on providing them with conditions which meet their needs (p. 49), and taking account of how they respond to particular features of their environment (p. 50–5). Seeds react to the presence or absence of light (p. 51), and are able to detect and respond to changes in its spectral quality (p. 53). Germination is almost always strongly affected by temperature (p. 53), including the natural fluctuations caused by night and day (p. 54).

The ways in which seeds are sown (p. 56), and handled when they have grown into seedlings (p. 58), are very important. They decisively influence the proportion of seeds which germinate, and will always have an effect on the rate at which seedlings develop (p. 60), and the time it takes them to grow into flowering plants.

Seeds can be sown out of doors, in shallow drills, and this is a method which requires very little equipment, and the minimum of care and attention. Annuals may be sown directly into the positions where they are intended to flower (p. 61). Seeds of biennials and longer-lived plants are usually sown in nursery beds (p. 31), and these can be used to produce large numbers, when needed, of hardy perennials (p. 102); shrubs (p. 122); conifers (p. 68); and trees (p. 149).

Protected conditions, provided by frames, greenhouses or propagators (p. 33), with or without the use of supplementary warmth from soil-heating cables (p. 36) or other sources of artificial heat, greatly increase the chances of success and the variety of plants which can be grown. Plants grown traditionally as annuals nearly all germinate with very little difficulty (p. 46). This quality is not confined to annuals, and plants which can be grown from seeds with equal ease are listed amongst alpines (p. 87); hardy perennials (p. 101); bulbous plants (p. 155); and shrubs (p. 122).

The seeds of some plants do not produce seedlings so easily, and may need quite complex sequences of events before they will germinate. Conditions which may require special treatments include: a short-lived inability to germinate when freshly shed (p. 103); requirements for conditioning treatments at low temperatures (p. 88); requirements for conditioning treatments at high temperatures (p. 104); the presence of immature embryos (p. 103); hard and impermeable seed coats (pp. 50 and 103); and the presence of chemical inhibitors which prevent germination (p. 123).

A variety of treatments can be used, separately or in combination, to produce seedlings from seeds which germinate less than easily. These include: chipping or fracturing hard seed coats (p. 50); conditioning at low (p. 88) or high (p. 103) temperatures; or stratifying seeds (p. 123), sometimes for periods of many months, to remove inhibitors, encourage the development of embryos or break down impermeable seed coats.

Ferns are grown from spores, which are not seeds, but a different stage in the plant's life cycle (p. 105). These must be sown and treated in appropriate ways (p. 108) before they will produce plants.

Section Two: Propagation from Cuttings

Cuttings are parts of a plant, separated from their parent and used to produce new plants. They share the genetic constitution of their parents, and almost always develop into precise replicas of them. Sets of plants propagated vegetatively in this way from a single mother-plant are known as clones.

Almost any part of a plant may be used as a cutting provided it contains cells which are capable of reproducing themselves by division and then differentiating to form new tissues (p. 7). Cells of this kind occur in characteristic positions. Any regenerative cell, irrespective of its origin, is able to reproduce all the new tissues and structures needed to make up a complete and fully functional plant. This phenomenon, known as totipotency (p. 8), is the cornerstone on which successful propagation from cuttings is based.

Cuttings are most frequently prepared from shoots, but roots (p. 111) and single buds (p. 138) may also be used as well as parts of leaves (p. 95). Specialised tissues from bulbs (p. 167), corms (p. 168) and rhizomes (p. 168) also contain regenerative tissues which make it possible to use sections of them as cuttings.

Cuttings must be kept alive and functioning while they regenerate the new tissues which enable them to resume independent function. They need protection to enable them to survive (p. 34), and conditions (p. 73) which allow them to function as normally as possible: these include cutting composts (p. 76) of the right kind, warmth, light and protection from desiccation (p. 73). Synthetic hormones (p. 74) may be a useful means of stimulating the production of roots.

Soon after roots or new shoots have been produced, the cuttings may require a short period of weaning (p. 82) before they become capable of renewed independent existence. They also depend on additional nutrients (p. 82) to support new growth and development, before being potted up individually (p. 82), and grown on in the protection of a greenhouse or cold frame. Later they may be planted out in a nursery bed (p. 83) or moved into a larger container, to complete their development.

Shoots may be cut in a variety of ways to produce basal, heel or nodal cuttings (p. 69). They can be removed at different times of the year, and at different stages of development. The growths used to produce cuttings include: very soft, immature shoots called tip cuttings (p. 128); shoots, in a semi-mature condition, just beginning to firm up, as the woody tissues within them develop (p. 130); the fully mature woody shoots of deciduous plants after the leaves have fallen (p. 133) or of evergreens during the winter (p. 135). The mature shoots of a wide range of semi-tender evergreen, silver-leaved or aromatic shrubs and perennials (p. 140) provide cuttings during the autumn which can be used as a reserve in case of losses during the winter.

Cuttings can be prepared from sections of stem, each containing a single bud (p. 138), as a convenient way to propagate climbers with lengthy internodes, or an economical way to use the shoots of roses (p. 138). During the summer, semi-mature shoots of many alpines (p. 91) or herbaceous perennials (p. 116) produce roots without difficulty and, in spring, cuttings made from basal shoots of herbaceous perennials (p. 110) develop quickly into new plants.

41

The root-like stems known as rhizomes can be used as a simple and trouble-free way to increase some alpines (p. 98), herbaceous perennials (p. 111) and bulbous plants (p. 168). Some alpines and herbaceous perennials may require modifications of normal techniques to propagate specialised shoots of various sorts including rosettes and offsets (p. 94) and leaf cuttings (p. 95) and leaf cuttings (p. 95).

Trees share the woody, persistent nature of shrubs, and many can be propagated from cuttings in similar ways. Tip cuttings may depend on special preliminary treatments to the parent stock plants (p. 152). But semi-mature (p. 152) and hardwood cuttings (p. 152) can be taken from suitable subjects using similar techniques.

Even bulbs (pp. 167), and many corms (p. 168), can be propagated by cutting them into sections and looking after them while they regenerate new tissues. Success depends on understanding the structure of these organs (p. 163) and the ways they relate to and resemble the shoots and rhizomes of more familiar plants.

Section Three: Division

Many plants reproduce vegetatively, under natural conditions, by producing ready-made replicas which are supported and protected by their parents while they become established and later develop into separate and independent individuals.

Herbaceous perennials produce numerous shoots, known as crowns, which gradually spread sideways; these can be divided and planted or potted up individually as a simple means of increase (p. 117). Many alpines behave similarly, or more aggressively (p. 97), producing mats of tangled stems which produce roots spontaneously as they spread over the surface of the ground.

Most bulbs (p. 169) increase naturally by sub-dividing and producing offsets, which provides a simple, though relatively slow, means of increase in gardens. Some corms (p. 170) form large numers of cormels around their bases, and these can be detached and grown on until they become large enough to produce flowers. Lilies (p. 170) produce bulbils along their flowering stems or just below ground level.

The stems of shrubs can be arched to the ground when many will form roots at the point of contact. These rooted branches known as layers (p. 143) can later be detached and planted elsewhere to produce new plants. Suitably placed side branches of trees (p. 153) can be treated in the same way. A technique known as air-layering (p. 146) provides a possible alternative means of propagating upright-growing trees and shrubs whose branches cannot be brought down to ground level. The trailing shoots of a small number of lax-growing shrubs like brambles (p. 147) form roots when their tips meet the ground, and this natural means of increase can be used to propagate them in gardens.

The roots of many shrubs and some trees spontaneously produce shoots at intervals along their length, which break through the ground as suckers (pp. 143, 153). These can be dug up with their roots attached to provide ready-made new plants.

Section Four: Collection and Storage of Seeds

Seeds are a convenient way to store plants for long periods if need be, with the minimum of effort, and in very little space.

The great majority of seeds, usually described as small, dry seeds, can survive intense desiccation, which reduces the water content of their tissues to 5 per cent or less of their total weight. These can all be stored for long periods easily, and very economically, using readily available equipment (p. 181).

A minority of seeds, mostly produced by trees (p. 180), cannot survive desiccation, and perish if the water content of their tissues falls below about 35 per cent. These include nuts and similar structures like acorns, and cannot be stored for longer than a few months; special precautions are often required to prevent deterioration within a matter of weeks (p. 181).

The ways that seeds are collected (p. 173) and dried after collection (p. 180) affect their quality, and the length of time they will remain alive during storage. Almost all seeds produced in fleshy fruits can be stored for long periods, but many require special treatment (p. 178) during and after the time they are harvested. Before they are stored they must be separated from all traces of the tissues of the fruits which surround them (p. 179).

Fern spores (p. 107) can be collected by drying their fronds over sheets of paper. Many can be stored under the same conditions as small dry seeds.

CHAPTER FIVE

Annuals and Biennials from Seed

Introduction

Most gardeners, when they think of annuals, think of summer bedding and the plants they sow in the spring, which are colourful throughout the summer, and fade away sometime before the onset of winter. Many do exactly that, but others do not, and in practice the definition is not quite so uncomplicated, and the times when plants grow, and flowers are produced, can depend very much on the part of the world in which they are.

Under natural conditions annuals growing around the Mediterranean Sea germinate during the autumn and produce flowers in the late spring; larkspurs and love-in-a-mist, amongst others, are sufficiently hardy to be grown in the same way in British gardens. The season when many of the annuals from tropical or sub-tropical parts of the world germinate and grow to maturity is dependent on when it rains, irrespective of the seasons, but they are confined to a summer-growing season in the British climate by their lack of tolerance to cold weather. A number of plants to be found in gardens, most from cool-temperate parts of the world, extend their life cycles over two growing seasons. Gardeners like to call these biennials, but, like annuals, once they have finished flowering they produce seeds and die, and the distinction between them and the plants we think of as true annuals is a very slender one. We usually sow biennials a little before mid-summer, to produce well-established plants which can survive the winter out of doors and come into flower the following spring; a pattern which bears a close resemblance to the normal behaviour of annuals in warmer climates like the Mediterranean.

Rather confusingly, a significant proportion of the 'annuals' grown in gardens are perennial plants in their native countries, but are grown as annuals in Britain because they are able to produce flowers quickly from seed sown in the spring, but are too tender to survive the winters. This group, which includes gazanias, petunias and lobelias, and nowadays the seed-raised strains of the zonal-leaved pelargoniums called geraniums, can all be propagated from cuttings, as well as grown from seed, provided they are protected during the winter. This is a way that particularly attractive forms can be kept going from one year to another.

Evidently, the annuals in our gardens are a mixed lot, and whether we label plants as annuals or not sometimes depends on how we have become accustomed to grow them. In this chapter annuals are not defined in a narrow botanical sense but the description is used to cover all sorts of plants used in that way in our gardens. They include a number which are not hardy, a category excluded elsewhere in this book, but included here because the plants concerned are well-established, colourful features in gardens.

Annual plants and biennials are monocarpic—they flower and then die, and are totally dependent on reproduction from seed for their survival. Their prospects of survival depend very critically on the season in which they germinate and seeds are naturally adapted to monitor the progression of the seasons, and to respond to them by producing seedlings at appropriate opportunities. In finer detail, they must be capable of assessing their depth of burial in the soil, the presence or absence of overshadowing vegetation, and other vitally important features of their environment such as the temperature, and the water content of the soil around them. When we sow seeds under artificial conditions in containers or a seed bed in the garden, we can expect them to germinate only:

if we provide them with all the components of their natural environment which are necessary for germination, in a form, or sequence, which sets off the chain of events which leads to the production of seedlings.

This may sound like a stuffily pedantic way to refer to the germination of seeds of mustard and cress, which need only a pad of moistened paper towel, and a warm corner by the kitchen window to persuade them to produce seedlings, and for many other seeds which germinate as easily. It becomes pointedly relevant when trying to unravel the responses of seeds of other more recalcitrant species, which can only be coaxed to produce seedlings by a series of events which must relate very closely to the conditions they would experience naturally.

The annuals and biennials traditionally grown by gardeners may be a bit of a hotchpotch, but with few exceptions they do oblige us by germinating when we sow them. Almost all can normally be expected to produce seedlings more or less straightaway, even those with more complex requirements, if the seeds are supplied with water and oxygen at favourable temperatures. In spite of their varied origins the annuals grown by man are a chosen few—out of the vast number available—and chosen inadvertently, but almost inevitably, just because they are accommodatingly easy to germinate. There is no reason to believe that the seeds of monocarpic plants, as a whole, are either less or more exacting in their germination requirements than seeds of perennial plants.

The Natural Distribution of Some Plants Grown as Annuals

The plants marked * in the list below are grown as annuals for summer bedding, but are perennial plants in their native lands, and can be grown as such if protected from frosts during the winter.

45

Plants from Areas with Mediterranean Climates

These parts of the world experience cool, moist winters and hot, dry summers. Almost all the plants from these areas can be sown *in situ* in the garden, or their seeds can be germinated and grown on at temperatures well below 15°C (59°F) in a greenhouse or cold frame.

From Southern Europe, North Africa and Western Asia

Antirrhinum majus
Borago officinalis
Brassica oleracea
  B. rapa
Calendula officinalis
Campanula media
Centaurea cyanus
  C. moschata
Cheiranthus cheirii
Chrysanthemum
carinatum
  C. coronarium
Convolvulus tricolor
Delphinium ajacis
  D. consolida

Dianthus barbatus
Echium plantagineum
Gypsophila elegans
Helichrysum orientale
Iberis umbellata
Lathyrus odoratus
Lavatera trimestris
Limonium sinuatum
  L. suworowii
Linaria maroccana
Linum grandiflorum
Lobularia maritima
Malope trifida
Matthiola incana

Moluccella laevis
Nigella damascena
Onopordum arabicum
Papaver somniferum
Reseda odorata
Salvia horminum
Satureia hortensis
Scabiosa atropurpureum
Sedum coeruleum
Senecio maritima*
Silene pendula
Silybum marianum
Viola wittrockiana
Xeranthemum annuum.

From California and Southern Oregon

Clarkia elegans
  C. grandiflora
Eschscholzia californica

Gilia capitata
Nemophila menziesii

Phacelia campanulata
Limnanthes douglasii.

From Western Australia

Ammobium alatum
Brachycome iberidifolia

Helichrysum bracteatum
Helipterum manglesii

Lobelia ilicifolia.

From South Africa

Anchusa capensis
Arctotis grandis
Dimorphotheca aurantiaca
Gazania hybrida*

Lobelia erinus*
Mesembryanthemum
criniflorum
Nemesia strumosa

Thunbergia alata
Ursinia anethoides
Venidium fastuosum.

From Chile

Nierembergia coerulea

Salpiglossis sinuata

Schizanthus pinnatus.

From the Tropics and Sub-tropics

Almost all these require comparatively high temperatures, between 15° and 25°C (59–77°F), for successful germination. Only a few can be grown satisfactorily without artificial heat, or sown *in situ* outdoors.

From Mexico and Central America

Ageratum houstonianum　*Ipomoea tricolor*　*Tithonia speciosa*
Cosmos bipinnatus　*Penstemon gloxinioides**　*Zinnia elegans.*
*Cuphea ignea**　*Tagetes erecta*
　*C. miniata**　　*T. patula*
*Dahlia variabilis**　　*T. tenuifolia*

From Tropical South America

*Alonsoa warscewiczii**　*Mirabilis jalapa**　*Portulaca grandiflora*
Amaranthus caudatus　*Nicandra physaloides*　*Salvia splendens*
Begonia semperflorens　*Nicotiana affinis**　*Tropaeolum majus*
Cleome spinosa　*Petunia hybrida**　*Verbena hybrida*.*
Gomphrena globosa

From Tropical Africa or Asia

Impatiens sultanii　*Ocimum basilicum.*

From Cool Temperate Regions

These include many biennials, which can be sown during the summer to flower the following year. Most germinate at moderate temperatures between 10° and 20°C (50–68°F) and can be sown outdoors in nursery beds or *in situ*, or germinated in a greenhouse.

From Eurasia

*Bellis perennis**　*Lunaria annua*　*Papaver rhoeas*
Digitalis purpurea　*Myosotis sylvatica*　*Viola wittrockiana.*
Kochia scoparia

From North America

Coreopsis tinctoria　*Helianthus annuus*　*Phlox drummondii*
Euphorbia marginata　*Oenothera biennis*　*Rudbeckia hirta.*
Gaillardia pulchella

From China

Althaea rosea　*Callistephus chinensis*　*Dianthus sinensis*.*

The Responses of Seeds to Their Environment

Seeds are inert only when they are dead, and living seeds, even those we put aside in paper envelopes in desk drawers, respond actively to the world around them. Once they have been sown, and their tissues become fully hydrated, these responses become by comparison hyperactive as the seeds monitor and react to their surroundings. Before it produces a seedling a seed needs to establish where it is, what is going on around it, and what prospects of survival the emerging

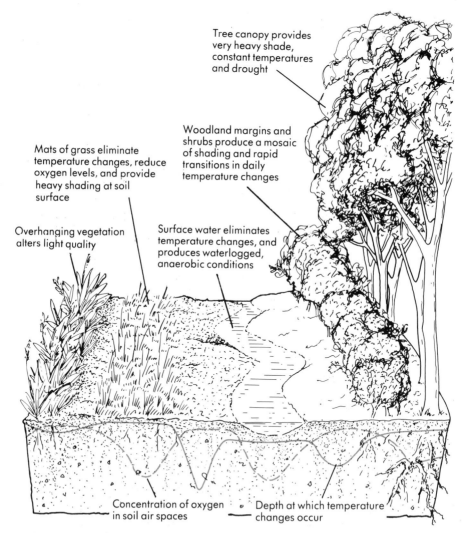

Tree canopy provides
very heavy shade,
constant temperatures
and drought

Woodland margins and
shrubs produce a mosaic
of shading and rapid
transitions in daily
temperature changes

Mats of grass eliminate
temperature changes, reduce
oxygen levels, and provide
heavy shading at soil
surface

Overhanging vegetation
alters light quality

Surface water eliminates
temperature changes, and
produces waterlogged,
anaerobic conditions

Concentration of oxygen
in soil air spaces

Depth at which temperature
changes occur

Figure 5.1   Seeds germinate in response to changes in temperature, the intensity and quality of the light that reaches them, and the amount of oxygen in the air spaces in the soil around them.

seedling would have. The critical features it surveys are the nature of the place where it lies, and the major physical features of the environment around it. It will be aware of and respond to light or dark, and discover whether the light which reaches it comes from a clear sky, or passes through overhanging foliage on the way. It will record whether the surrounding soil is cool or warm, or perhaps freezing cold; the presence of carbon dioxide and the 'freshness' of the air; whether the ground around it is wet, or moist, or dust dry, and whether it remains consistently so or varies from day to day. It will assess the fertility of its position and may react to the presence of unusual salts or other chemicals.

Characteristics of Seed Composts

Gardeners usually, though not invariably, sow seeds in a peat- or loam-based compost of one kind or another. Seeds will germinate satisfactorily in a great variety of substrates or composts, but it simplifies management and makes success more probable if they possess certain qualities. A good substrate should be:

- Open enough, in structure, to allow emerging roots and shoots to move easily, but dense enough to support seedlings.
- Sufficiently stable to remain in good condition long enough for the seeds to produce seedlings.
- Capable of holding and retaining high water levels without becoming waterlogged.
- Sufficiently porous for oxygen to diffuse freely even when saturated with water.
- Easily rewetted after drying out.
- Transparent, rather than opaque, so that light reaches seeds buried beneath the surface.

Responses to Water

The presence or absence of water is far and away the most important single feature which decides whether a seed will germinate or not. The amount of water in a seed varies tremendously, depending on where a seed is and how moist or dry its surroundings are. Seeds kept in paper packets, more or less dry in a drawer, will take up or lose water from day to day as the atmosphere around them varies in humidity, but usually between 5 and 25 per cent of their total weight will be water. There are some seeds, including most nuts, which die if they dry out to this extent, and these survive only while their tissues retain 60 per cent and more of water. This difference in tolerance to desiccation between the two groups of seeds becomes very important when attempts are made to store them for long periods. Seeds which can be dried without harm are much easier to keep alive (p. 181), if need be for many years, than those which shrivel and die if they lose too much water.

Water passes with very little hindrance through the seed coats of most seeds, either in or out, and exposure to water in the soil, or a humid atmosphere, immediately results in water entering the seed in a process known as imbibition. The water content of fully imbibed, water-saturated seeds can be as high as 70 or 80 per cent of their total weight, and at these levels most seeds become capable of germination.

Seeds soak up water voraciously, and are able to hold many times their weight of water. However, the fact that a seed takes up water and swells does not necessarily mean that it is still alive. Inadequate or spasmodic watering, which results in alternate wetting and drying of seeds, or partial hydration, can be lethal, especially if the seeds happen to dry out after seedlings have started to germinate. When attempting to germinate seeds one of the most important things is to make sure that abundant free water is available throughout the entire period from the time the seeds are sown until seedlings can be seen emerging from them. The simplest way to do this is to use free-draining composts which can be watered heavily without becoming waterlogged.

Most seeds imbibe water spontaneously whenever they come into contact with it, but it is not uncommon to come across seeds, members of the pea family are one example, which are waterproofed by the presence of fats, or waxes, in their outer covering, and in which the embryo is unable to take up water at all until the seed coat is fractured, or damaged in some way, so that water can penetrate it. Usually, seeds of this kind can easily be recognised because their seeds fail to swell when watered. The simplest remedy is to chip away a small portion of the seed coat, to allow water to reach the interior of the seed.

The sudden removal of part of the seed coat destroys the main means by which seeds control the uptake of water and the rehydration of their tissues. Sometimes seeds treated in this way fail to produce seedlings after they are sown and watered, but go mouldy and rot. The most probable reason for this failure is that the inner parts of the seed are exposed to the effects of water before they are sufficiently organised to cope with it. This can be avoided by allowing the membranes time to repair themselves after chipping the seed coats, and before they are sown and watered. Normally, they will do this if left in the open air for 24 hours: small amounts of moisture absorbed from the atmosphere enables them to resume limited activity and repair their defective membranes. Later when the seeds are sown and watered these are again fully functional and ready to resume their control of the imbibition of water in an orderly way.

Effects of Oxygen and Carbon Dioxide

Oxygen is as necessary to plants as is the air we breathe to us, to provide for the processes of respiration by which storage reserves and other sources of energy are used to support growth and day-to-day existence. Seeds resting in the soil tick over so minutely slowly that they require exceedingly little oxygen. Germination, by comparison, is a time of great activity when reserves are broken down to support rapid growth, and most seeds then need much more oxygen to function effectively, and depend on its ready availability. Usually the levels of oxygen found in the air, and diffusing into the surface layers of loose soil, are sufficient, and examples of situations where germination has been improved by giving seeds extra oxygen are more academic than practical. However, sub-normal levels of oxygen can reduce germination, and these are by no means particularly uncommon or at all academic. They occur in waterlogged composts; they may result from sowing seeds too deeply; or develop if the surface of the compost above the seeds becomes compacted, or clogged with algae or mosses. The seed's needs for water can conflict with its need for oxygen, since water can fill all the available spaces in a compost causing waterlogging and a dearth of oxygen. The capacity to provide both water and oxygen is an essential quality of a good compost or seed bed, and one of the most important features which governs its composition.

Carbon dioxide is produced during respiration, and, when reasonable ventilation prevails, plays little or no further part in the ways that seeds germinate. If seeds are buried too deeply, or the soil above them is compacted or impermeable, carbon dioxide, produced by the respiration of the soil flora and fauna, builds up within the deep recesses in the ground and, at these higher than normal levels, has a narcotic effect which can reduce activity within the seeds, and make their germination less probable. The balance between oxygen and carbon dioxide which they experience is one of several ways by which seeds assess how deeply they are buried, and whether an emerging seedling would have a reasonable chance of struggling up to the surface and into the sunlight, without which it could not survive.

Perception of Light

Light provides seeds with vitally important information about the time of year, and the conditions that seedlings would encounter if they emerged. Many, if not most, seeds can detect very low levels of light, and discriminate accurately between critically important wavelengths within the visible part of the electromagnetic spectrum. The ways in which seeds respond to light are often strongly affected by temperature, and these two features of the environment interact in complex ways, which do nothing to make the gardener's lot easier.

The seeds of a great many plants germinate adequately whether in light or in darkness, and the need for one or the other has little practical importance. There are many others, however, which depend on a period of illumination for at least a part of each day and in continuous, total darkness they either produce no seedlings at all, or the numbers they produce are very small. A tiny minority of plants have developed seeds which germinate only when they are in total darkness, or do so more rapidly than when they receive a daily measure of light. These reactions to light and dark leave the gardener in a dilemma. The appearance of a seed provides no clues to the ways it will react, nor do any particular features of the plant's life style or origins. Obviously the answer is to look up a list, obligingly prepared by someone else, but such lists are exceedingly misleading, often being based on a naïvely simple approach to responses which are not at all clear-cut, but are complicated by interactions with other features of the environment, particularly temperature.

Different varieties, even of one kind of plant, vary one from another in the degree to which they depend on light for germination. There are innumerable examples, such as lettuce, where a particular cultivar of a plant depends on light, and another is indifferent; there are many others where, at one temperature, seeds must be illuminated each day before they will germinate, but, when they are slightly warmer or perhaps cooler, germinate when in continuous darkness. The gardener's most prudent course is to accept that the overwhelming majority of seeds either require light at least under some conditions, or are not affected by it one way or the other. The number that are known with certainty

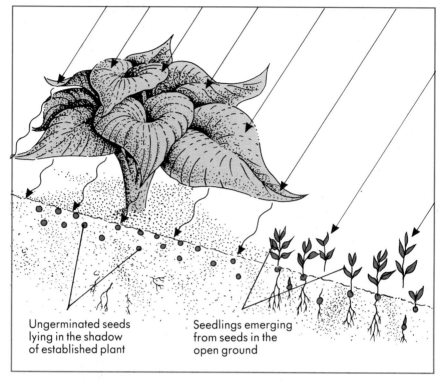

Figure 5.2   Overshadowing leaves alter the quality of the light reaching seeds lying in their shadows, and may prevent them from germinating.

to depend on continuous darkness is very, very small, and some of these owe their reputation to experiments done, many years ago, with light sources whose curious spectral compositions produced misleading results. The best answer is to make certain that all seeds are sown in ways which ensure that they receive some light each day: unless there is a very strong reason to believe they are one of the small number which really do germinate more successfully in total darkness.

Seeds are most easily deprived of light by sowing them too deeply in a light-absorbing compost, and of all the composts used, peat is the one which most effectively absorbs and limits the penetration of light. This reaction exactly mimics the response of many seeds under natural conditions, by which the presence, absence or intensity of light provides an indication of the depth of burial beneath the surface of the soil. Occasionally, suggestions are made that seed pans should be covered, after sowing, with totally opaque materials like slates, but such coverings are better avoided in favour of substances which will let some light pass through them.

Traces of light may be all a seed needs to tell it something about its depth of burial in the soil. They cannot warn it that any seedling produced would emerge immediately beneath a canopy of leaves from neighbouring plants which would ensure its early demise. Leaves are transparent to light and seeds lying beneath even a dense canopy would receive quite sufficient illumination to trigger off the responses which produce seedlings. However, as light passes through leaves it is partially absorbed, and emerges on the other side not only paler than it went in, but different. Some wavebands are almost totally absorbed, others pass through with little or no reduction.

Plants perceive light only in a few critically important wavebands, unlike our eyes which respond to a comparatively broad range of the spectrum, but seeds are able to compare the relative intensities of radiation in particular wavebands, so that they are good at 'seeing' quite small differences in the colour of light. This ability makes it possible for them to discriminate between light which has been changed by passing through an overhanging leaf, and simple variations in the intensity of natural daylight, much more effectively than we can, so that they do not produce seedlings in situations where established vegetation would destroy any hopes of their survival. Containers in which seeds are sown should never be put in the shelter of other plants whose leaves might overhang them, and inhibit the production of seedlings. There is a slight possibility that coloured sheets of glass or plastic might absorb and transmit light in ways which affect germination, and these should be avoided by using colourless or white materials which do not distort the spectral composition of the light which reaches the seeds.

Effects of Temperature on Seed Germination

Gardeners may find cold fingers a trial during the winter, and look for shade to keep cool on a summer day, but, because they are warm-blooded mammals, they maintain a near-constant body temperature. Plants and seeds experience temperatures as they are in the air or ground around them, rising and falling in exact synchrony, so that temperature becomes a very important part of their environment and one which controls many of the ways that they grow and develop.

Seeds of different kinds of plants are able to germinate at temperatures which range from just below freezing point, beyond which they become too cold to function, to about 42°C (108°F), which is the highest temperature that most hydrated seeds can endure for more than a few days without dying. Few it any species are capable of producing seedlings over the whole of this range, but different kinds do so over very distinct temperature spans, and laboratory experiments have shown that these ranges are often extremely closely defined. Although changes in temperature of only 1 or 2°C (2 to 4°F) may reduce or increase germination by 50 per cent or even more in a laboratory, they have much less noticeable effects under practical conditions.

In most greenhouses and cold frames, let alone out of doors in seed beds, variations in temperature occur naturally which swamp small differences that turn out to have such large effects in experiments. But, when thermostatically controlled propagators, or well-constructed, well-controlled installations using soil-heating cables are being used, the effects of these small changes in temperature are much more likely to become relevant. It is easy to achieve conditions which constantly maintain temperatures above the maximum at which germination occurs.

A very large proportion of the hardy annuals which we grow in gardens come from temperate parts of the world and are able to germinate rapidly at temperatures between 10 and 20°C (50–68°F). Many will germinate in time at far lower temperatures, and some, the corncockle for example, are capable of eventually producing seedlings even when they are embedded in ice. But, unless time is pressingly important, there is no other advantage in using higher temperatures. An unheated greenhouse or cold frame can often give satisfactory results, even though they may take longer to achieve than they would in a heated greenhouse or propagator, provided seeds are not sown too early in the year. The temperatures referred to may seem low in comparison with those usually quoted for the seeds of bedding plants in particular. This is mainly because little or no distinction is made between the needs of seeds of plants from sub-tropical and tropical regions, which almost all require high temperatures to ensure good results, and seeds from temperate regions with much more modest needs.

When gardeners start to control the conditions in which they grow their plants, they sometimes develop an inclination to control things as closely as possible. So, if a greenhouse is used to provide a hot, moist atmosphere it is managed so that as far as possible it remains constantly at exactly the same levels of humidity and warmth, and deviations are looked on as shortcomings. Similarly, thermostats on electric propagators are set to provide constant temperatures day and night. These are very unnatural situations, particularly in temperate parts of the world where changing conditions from day to day, regular variations from day to night and very marked seasonal characteristics are normal. Very many seeds depend on seasonal changes to produce the conditions under which their seeds germinate, and an important minority of plants produce seeds which germinate better, at least under certain circumstances, when they experience the natural fluctuations in temperature between day and night, and there are some which do not produce seedlings at all at constant temperatures. This sensitivity to fluctuating temperatures is yet another of the ways by which seeds assess their depth of burial in the soil. Seeds close to the surface are likely to feel the chill of night, and the warmth of the sun by day; those too deep down, far below the surface, from which no seedling could emerge, exist in impenetrable darkness surrounded by cool earth, which does not respond to short-lived changes in temperature between day and night.

The ways in which seeds respond to temperatures which go up and down between day and night are academically interesting, but also affect how we do things practically. A plant as familiar to us all as celery can fail to germinate in temperatures which remain unchanged, or may do so only slowly and in inadequate quantities at low temperatures. Some strains of snapdragon produce seedlings only when alternating temperatures persuade them that day and night still exist. There are very few seeds which depend for germination on constant temperatures, and as a matter of common prudence, seeds should always be sown in circumstances which ensure that normal fluctuations from day to night prevail, at least till the time when seedlings emerge. It may be better to abandon smug satisfaction with the constant temperatures provided by recently installed electronic devices and turn down the heat at night, or switch it off and save heat and money. Seeds in small containers clustered around the hot-water tank in the airing cupboard may benefit from a daily respite, standing on a window sill.

Effects of Soil Fertility on Seed Germination

Successful germination seldom depends on the fertility of the compost, but may sometimes be reduced by highly fertile conditions, and especially by the presence in the surrounding soil of high concentrations of inorganic salts or other chemical compounds. Traditionally, seed composts contain smaller quantities of added fertilisers than potting composts, but this is not a practical necessity for germinating the seeds of most annuals. Nurseries, where very large quantities of compost are used, can make worthwhile economies by reducing the fertilisers added to the composts in which seeds are sown. On a smaller scale the use of two different composts, one for sowing seeds, and the other for growing on seedlings, is itself likely to be a source of waste, and almost always it would be more economical either to use a dual-purpose compost, or to use a standard potting compost for both purposes.

Sowing Seeds and Pricking out Seedlings

Traditionally, seeds are sown by scattering them over the surface of a suitable compost, and then covering them with a layer of sifted compost, applied very scantily, or more thickly, according to the size of the seed sown. The method has been used very successfully by gardeners for many centuries, and there can be no doubt that it has stood the test of time. Nevertheless, it does not always produce good results, and often fails to provide seeds with the best possible conditions in which to germinate and develop into healthy seedlings.

The compost used, whether based on peat or a fibrous loam, is itself the cause of some of these problems. The physical qualities required to provide good drainage, high water-holding capacity, good porosity to encourage oxygen diffusion, and the stability needed to maintain these features, are most easily obtained from a compost with a high proportion of coarse, fibrous particles in its composition. A coarse-grained compost of this kind is not so good for seeds while they are imbibing water, and for very young seedlings establishing the first traces of a root system. The usual solution to this problem is to use a fibrous, relatively coarse compost as a base; to level its surface, and perhaps top it with a light layer of sifted compost before sowing seeds; and finally to cover the seeds themselves with a light topping of sifted compost.

The seeds are sown in a layer of very fine compost from which all the large and more fibrous particles have been removed, and this is the part which is most likely to cause problems. If water is applied too vigorously the seeds in it become water-borne and start to float—most will end up in a pile in one corner of the container. It is not easy to apply the final covering over the seeds evenly, or even to judge its exact depth, without a good deal of experience and confidence. The layer of dust-like particles is not at all stable: loam-based composts tend to dissolve into an ooze of mud, or silt; peat-based composts to become dust-dry very rapidly or, if overwatered, to become slimy, the surface clogged with minute algae. Seedlings emerging in this compact layer, liable to drain poorly, become susceptible to stress, vulnerable to invasion by soil-born fungi and may start to damp off. These traditional ways of sowing seeds are not easy for part-time gardeners to manage; small errors or misjudgements quickly lead to problems, even under ideal conditions, and when conditions are less than ideal—long spells of damp and low light in winter, bright sunshine and high temperatures in summer, or seeds which delay their germination, perhaps for weeks—it can tax the skills even of experienced gardeners to produce a satisfactory stand of seedlings.

Standard Method of Sowing Seeds

These problems can be avoided by sowing seeds in ways which are much easier to manage. The simplest method is based on the principle that a seed compost should consist of two parts: a lower layer which provides space for the roots to develop, and a reservoir of water and nutrients to support their growth, and an upper layer to accommodate the seeds and provide them with ideal conditions in which they can germinate, and seedlings can become established. Any good quality potting compost can be used for the lower layer, but this should be free-draining, and contain high proportions of fibre to enable it to hold water like a sponge, and yet drain freely to dispose of surplus water. The upper layer must consist of a material which is porous and water-retentive, but free-draining and very stable. Loams and peats do not match these requirements and are unsuitable in any form. Grits and sands dry out too rapidly and are not very satisfactory. An ideal material is a porous grit (similar in consistency to the debris of heavily crushed clay flower pots). Crushed clay pots have been included in recipes for potting composts from time to time in the past but are not available to many of us today. Alternative materials which are easier to obtain include:

(a) Calcined clay minerals of the kind that are used for cat litter, or soaking up oil spillages from garage floors, or marketed from time to time for horticultural use.
(b) Horticultural vermiculite or perlite.
(c) Crushed brick, marketed as a dressing for hard tennis courts.
(d) Crushed tufa, sifted to remove dust.

All of these have the right consistency, they are stable materials which retain their integrity for a long time and are easy to manage, they are inert, they hold water and yet drain freely, and produce an environment for seeds which is moist, well-aerated, and easily penetrated by the emerging plumules and developing roots.

In practice, containers in which seeds are to be sown are prepared by two-thirds filling them with potting compost, to provide the lower layer. The topping of porous grit or vermiculite is then added to fill the container, apart from a gap below the rim about 0.5 cm ($\frac{1}{4}$ in) deep to allow for watering. Very small seeds, such as lobelias and begonias, are scattered over the surface and allowed to sink into the upper layers of the grit when they are watered;

Figure 5.3   Seeds should always be sown in the smallest container that will hold the number of plants needed. A gritty well-drained compost reduces risks of damping off, and makes management easier.

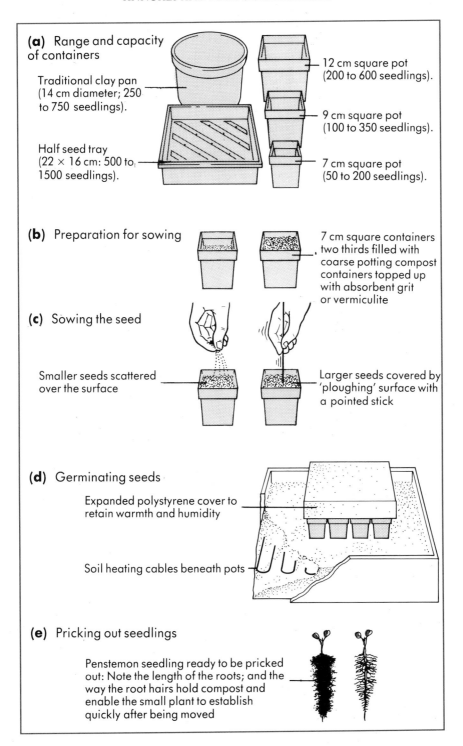

(a) Range and capacity of containers

Traditional clay pan (14 cm diameter; 250 to 750 seedlings).

Half seed tray (22 × 16 cm: 500 to 1500 seedlings).

12 cm square pot (200 to 600 seedlings).

9 cm square pot (100 to 350 seedlings).

7 cm square pot (50 to 200 seedlings).

(b) Preparation for sowing

7 cm square containers two thirds filled with coarse potting compost containers topped up with absorbent grit or vermiculite

(c) Sowing the seed

Smaller seeds scattered over the surface

Larger seeds covered by 'ploughing' surface with a pointed stick

(d) Germinating seeds

Expanded polystyrene cover to retain warmth and humidity

Soil heating cables beneath pots

(e) Pricking out seedlings

Penstemon seedling ready to be pricked out: Note the length of the roots; and the way the root hairs hold compost and enable the small plant to establish quickly after being moved medium-sized seeds, which would include pansies and cabbages, are scattered evenly over the surface, and then 'ploughed' in using a pencil or pointed stick to break up the surface and bury the seeds beneath it. Sweet peas, lupins and other larger seeds are sown on a shallow bed of the porous grit immediately above the lower layer of compost, and more grit is added to bury them about 1.5 cm ($\frac{1}{2}$ in) beneath the surface.

When seeds have been sown in this way their management becomes extremely simple. They should be covered until they start to germinate, and nothing does this better than sheets of expanded polystyrene—ceiling tiles can be used, the pot holders used to deliver pot-plants to florists shops, or sheets of salvaged packing material—all can easily be cut to exactly the size required, and the material does not provide a cold surface on which water condenses and drips on to the seeds below. When seedlings emerge these sheets should be removed each morning and replaced each evening, until the seedlings grow up and come into contact with the polystyrene. The seeds and young seedlings must never be allowed to become dry, or suffer stress from lack of water. If in any doubt they should always be watered copiously, using a watering can with a fine rose, from above. There is little danger of causing damage from overwatering because the layer in which the seeds are sown drains freely and retains no surplus water. Later, as seedlings develop, the parts which are most vulnerable to the fungal infection which cause them to damp off are also located in the porous, free-draining upper layer and are much less likely to suffer from these problems, even with less than skilful watering.

It is well worth making an attempt to standardise the containers in which seeds are sown if this is at all possible, by choosing a particular shape and size, appropriate to the number of seedlings needed. Plastic containers are easier to manage than clay ones, and a square cross section is more economical than a round one. Drainage is vitally important, both through the compost surrounding the seeds and out of the base of the container, because much the easiest system of management is to water frequently and allow surplus water to drain away. In normal circumstances the smallest square section plastic container, which will accommodate sufficient seedlings, is the right one to use. A small square plastic pot with a surface area of about 50 sq cm (i.e. 7 × 7 cm) ($7\frac{1}{2}$ sq in, $2\frac{3}{4} \times 2\frac{3}{4}$) will hold about 100 seedlings of all the more finely built bedding plants, up to the time they are ready to be pricked out, about 50 African marigolds and something like 25 seedling dahlias. Fifteen of these pots will fit into the space occupied by a single seed tray and the latter, which is constructed to hold thousands of seedlings at this stage of their development, is almost always wastefully large for the purpose.

Pricking Out Seedlings

Traditionally, whether plastic pots or seed trays are being used, seeds are sown quite thickly, but not too thickly, and soon after they have germinated the seedlings are transplanted into another container, individually or in small groups, by an operation known as pricking out. At this stage each seedling is spaced out to give it room to develop and a seed tray becomes an appropriate, convenient and practical choice. Seedlings should be pricked out at the first moment when they can possibly be handled. Normally this is when the seed leaves, known as cotyledons, have just become fully expanded—a prospect which dismays many novice gardeners. It appears to be impossible to handle something so small and certainly absolutely impossible to handle it without crushing it to death. In fact these tiny seedlings are surprisingly easy to handle, holding them firmly by their newly expanded seed leaves, and they suffer much less damage than they would if attempts were made to move them later when they appear to be more robust. Many seedlings produce roots which grow so rapidly and strongly that any delay results in damage to the roots and makes it much more difficult to place the seedlings quickly and neatly into their new positions, using a small pointed stick to open up planting holes and guiding each seedling into it by hand.

The aim should be to keep these little plants in seed trays only for as long as they need to grow large enough to establish themselves successfully when planted out in their final positions in the garden. Wide spacings between seedings pricked out into seed trays waste space and are not helpful. A standard seed tray should hold about 40 plants of strong-growing annuals like dahlias; moderate growers like snapdragons and verbena about 70, and small tufty plants like alyssum and lobelia can be packed in so that each tray holds about 90 clusters of small plants. Seed trays are large objects which occupy a great deal of valuable protected space in springtime, and lap up enormous quantities of expensive compost, but too many gardeners are not aware of their potential capacity, and are reluctant to pack them with as many plants as they can hold.

Space Sowing Seeds

Sowing seeds thickly, and then pricking out the seedlings, saves space while the seeds are germinating, and allows for flexibility especially when variable numbers of seeds germinate in different batches; but pricking out seedlings is a chore and can be a major bottle-neck. It is all too easy to set to and sow all the bedding on one day, with a comforting feeling of a job well done only to find, three weeks later, that it is impossible to cope with them all and many have to be left hanging around and deteriorating while they wait their turn. One alternative, apart from planning things better, is to sow the seeds, a few at a time, straight into individual small containers, or spaced out in their positions in a seed tray. Spaced sowing avoids hold-ups while seedlings wait to be pricked out, but it occupies a great deal more space during early stages, and unless very high proportions of seeds germinate, so that every container or position holds at least one seedling, it can be exceedingly wasteful. It is worth contemplating only if good quality seed is available, and freshly purchased seed enclosed in foil is usually satisfactory, at least for vegetables; flower seeds sometimes fail to perform as well as might be expected however much care is taken to use recently purchased seeds. Older seeds, as well as seeds collected from the garden, and not carefully looked after, are always suspect and very seldom worth sowing in this way.

The advantages of spaced sowings often look very attractive at first glance, but they can be a little illusory. Although pricking out may not be necessary, thinning out clumps of seedlings sown together in each position can take almost as long. This is a tedious, fiddly task, especially if the seedlings are small with slender stems, and even more so if the sower has been impatient and heavy handed, so that great care and some skill is required to avoid damaging those seedlings that are left to grow on. There is a way out of the problem, although it will offend perfectionists, but which does take advantage of nature's tendency to solve problems economically and effectively. That is to leave the clusters unthinned, and let the stronger seedlings prevail and the weaker ones go under. This is what happens to annuals, growing under natural conditions, which frequently emerge in quantities far in excess of the numbers which could eventually survive. This appears to be wanton waste, but in reality provides a well-developed strategy which favours the survival of the species concerned. Large numbers of seedlings occupy space, and crowd out competition from other species; they also reserve room for whichever individuals amongst them survive the struggle for supremacy.

Times to Sow Seeds

Plants which are growing rapidly and healthily under favourable conditions are seldom difficult to manage or to grow successfully. But when struggling with poor light and cold whether they can be extremely demanding to grow thriftily, free from insect pests or fungal attacks. This statement of the obvious would not be worth making except that it is so frequently ignored as one of the most valuable guides to the management of plants. Most gardeners, free from the additional problems of having to find a market for the plants they grow, can choose the seasons when they sow seeds and commit themselves to producing plants.

But, very often this choice slips out of sight, as gardeners succumb to the annual temptation to use their greenhouses and electrically heated propagators: year by year annuals are sown needlessly early in the season, when artificial heat must be provided for long periods to maintain growth, and natural levels of light are low. Temperatures during the night almost always fall far below the levels that optimistic assessments during the day suggest, and the plants, suffering from bad conditions, damp off or fail to thrive. If, at great expense on artificial heating and lighting, they are persuaded to grow well they quickly become over-large, and suffer from being crowded together while it is still too chilly to move them out into a cold frame, let alone plant them outside.

The rates at which annuals developed when sown in a greenhouse at monthly intervals between 1st February and 1st May are shown in the table.

Spring stirrings that urge us to sow annuals in February, even in March, should be ignored under almost all circumstances. April is early enough and May not too late for almost all annual bedding plants. Late-sown plantings

Dates on which successive sowings were:				
Sown	Pricked Out	Planted Out	In Bloom	No. Days*
1 February	1 March	1 May	23 June	143
1 March	25 March	10 May	27 June	119
1 April	15 April	27 May	8 July	99
1 May	12 May	14 June	15 July	76

* Number of days from sowing seeds to flowering.

catch up so rapidly on early ones that plants sown in May and grown under good conditions without any setbacks will flower only a very few weeks later, and in half the time and a quarter of the expense of those started in February, and will more than make up for any delayed start by continuing to produce flowers long after the earlier sowings have gone to seed. When seeds have to be sown over as long a period as possible, for the very good reason that there is no other way to cope with the work, then the earlier sowings should be confined to annuals of Mediterranean origin (p. 46). These will accept low temperatures and poor light as part of their natural lot and take them in their stride. Annuals from warmer tropical climates which sulk unless pampered should be left till last.

Dealing with Pests and Diseases

Fungal diseases can be a problem, and overwhelmingly these will be the soil-borne, damping off diseases that can decimate seedlings of some annuals. Their presence should always be recognised as a visitation on the grower's mismanagement. They are encouraged by the use of unsuitable composts, poor growing conditions, crowded seedlings and lack of hygiene, and are almost always avoidable evils. Free-draining, sterilised composts, clean water, and an even, open pattern of seedlings help prevent the onset of infection: pricking out correctly—not too late, and neither too deep nor too shallow—growing on rapidly with adequate warmth, high light levels and plenty of ventilation are usually sufficient to stop them gaining a hold. The commonest source of infection is that much loved greenhouse installation the water butt; whether it comes in coloured plastic, as a folksy wooden barrel or just functional, zinc-clad steel, it is an ever-present refuge for these water-borne fungal diseases and a menace. Tap water, even if it does produce an occasional whiff of chlorine, is a much better bet.

Insect pests almost always become a problem only on problem plants, and seldom threaten the well-being of young, vigorously growing annuals. Colonies of young greenfly may set themselves up on young shoots and can be rubbed out between finger and thumb as soon as they are noticed. This is most effective, and environmentally benign, and similar methods of hand picking are practically the only way to deal with caterpillars. Many of these feed at night in company with slugs, snails and woodlice and an occasional foray with a torch after dark provides an opportunity to satisfy atavistic hunting instincts and keep the greenhouse free of pests. A comprehensive programme of spraying with insecticides, and, to be on the safe side, fungicides, backed up by regular blasts of toxic smokes, may seem more up to date, more in accord with twentieth-century practices. It will reduce the numbers of some harmful pests, and have little or no effect on the numbers of others, but it will also eliminate the predatory mites, ladybirds, anthocorids and lacewings which feed on and restrain the abundance of harmful pest species. The beneficial effects of using insecticides are likely to be more than offset by the loss of these predators, leaving the plants even more vulnerable to attack later on when the next generation of bugs arrives for its helping.

Annual bedding plants are grown in greenhouses for only a short time, at a stage in their development when they are young and vigorous and should not be very susceptible to infestations by insects and mites. It should seldom be necessary to be forced to take very severe measures against such pests, but if it is, the answer is to find out what has gone wrong, and put that right rather than use poisonous chemicals in attempts to treat the symptoms. More often than not the trouble will be traced to attempts to grow the plants under unfavourable conditions—too chilly perhaps, or with insufficient light—most probably due to sowing them too early, or using a badly sited greenhouse.

Direct Sowing of Annuals in their Flowering Positions

Many annuals, and those listed at the beginning of this chapter which come from Mediterranean areas are particularly suitable, can be sown *in situ* in the places where they are intended to flower out of doors. It is so easy to become accustomed to thinking of bedding plants as things that come out of seed trays from the garden centre, that the opportunities for growing them in this simple way are often forgotten. It can be a successful and effective way to grow them—it can be a total disaster. Success depends on the site, the weather and the nature of the soil, but unusually wet or dry conditions immediately after sowing pose problems against which little can be done. Success is most likely when it is possible to provide:

- A naturally friable soil which is easily broken down to a fine tilth to form a seed bed.
- A water-retentive, but well-drained soil, which does not dry out very rapidly, and can be watered, if need be, without forming a surface crust.
- A site which is warm and sheltered, especially for autumn sowings, but not in a frost pocket.
- A site completely free from perennial weeds, and not too heavily infested with the seeds of annual weeds.
- An open sunny position, not shaded by neighbouring trees, walls or buildings, and clear of the roots of neighbouring hedges.
- A stock of good quality seed capable of producing large numbers of vigorous seedlings.

The surface of the ground should be lightly forked over and raked to produce a seed bed immediately before sowing the seed, but artificial fertilisers need be added only when annuals are being grown on soils which are very infertile— thin sands and gravels or the threadbare remnants of organic matter that blacken the soil in long-deprived gardens in towns. Annuals are opportunists able to scrape a living under very poor dry conditions, in which even tiny plants will produce a flower or two and set what seed they can, but when overfed, they wax coarse and fat, with soft stems, abundant foliage, and a long wait before they produce flowers. The trick is to find a happy mean and make sure that the plants are adequately supplied with potash and phosphates, but kept low on the plant's equivalent of polysaturated fats—nitrogen. Light dressings of fish meal or bone meal will supply what they need, and encourage them to flower early and keep going through the summer.

Figure 5.4   Many annuals can be sown straight into the garden in the places where they will flower, but success depends on preparing the ground well, and care with sowing and weeding.

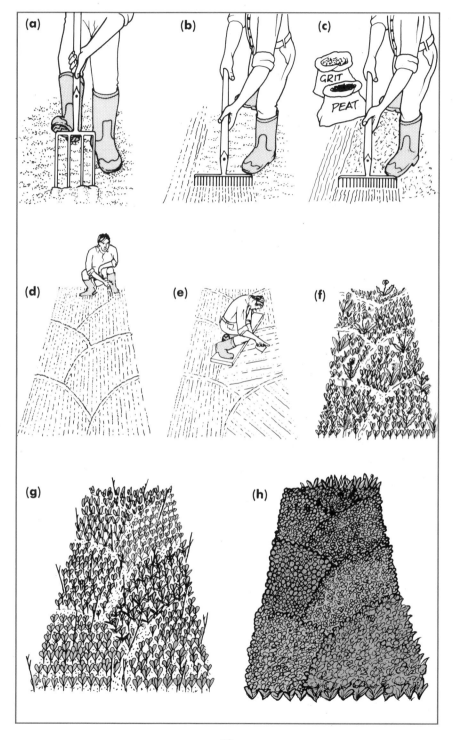

With the ground prepared and a tilth produced, all is ready for the seeds to be sown: the only thing that remains is to decide where different kinds should go, and mark out their positions on the ground. The most effective, simple way to do this is to use pale coloured sand or grit in thin trickles to mark out the boundaries of the territory to be occupied by each variety. The seeds can be sown by scattering them evenly, broadcast within their boundaries, but this is a very bold gesture. One of the biggest problems, and certainly the most fiddly operation met with when growing annuals in this way, is separating seedling weeds from those of the annuals, and this is a great deal more difficult in the hurly-burly that results from broadcasting seeds, than when they are sown in orderly lines. It is very advisable to sow each kind of plant in short straight drills within its territory. The straight lines provide easy guides to follow during the early stages when weeding is necessary, but do not remain defined for long, and soon disappear as the plants grow.

The nail-biting period starts after the seeds have been sown and, at this stage, a spell of really wet weather, especially on clay soils, can create a garden disaster which becomes a salvage operation; seeking seedlings amongst a mass of weeds. Under more favourable conditions, the ground should be gone over, soon after the seeds germinate, to identify the rows, remove weeds and transplant seedlings from places where they have come up too thickly to places which are a little threadbare. This can be followed up a few weeks later by another weeding, which with any luck will be the last, accompanied by ferocious thinning of the annuals. If left too close together, each will produce a small plant, with few side stems, a brief display of flowers and no follow up to extend the season. Widely thinned plants expand to fill the space available, and develop side branches which not only carry successive flushes of flowers, but themselves go on to produce yet more shoots which flower for months on end.

It is far easier to sow annuals in their flowering positions in this way when all is done on a small scale rather than as part of a grand design. A prudent start might be to sow small groups amongst perennials and shrubs, to brighten up a mixed border, or to furnish a narrow bed beside a path. Once this has worked successfully a few times, larger, more ambitious schemes become a reasonable ambition, but this is one of those gardening operations where problems multiply with complexity. Attempts to cover largish areas of ground with a varied tapestry of numerous multi-coloured annuals, provide so many hostages to fortune that they very rarely live up to their creator's dreams.

Propagating 'Annuals' from Cuttings

A sizeable minority of the plants we grow in gardens as annuals are not true annuals at all but tender perennials. These plants are identified by asterisks in the lists at the beginning of the chapter. Some of these, like the zonal pelargoniums and gazanias, are habitually raised from cuttings, but many of the others are now seldom propagated in this way. Cuttings of most of them, prepared and treated in the ways described in later chapters (p. 127), provide a way of perpetuating selected varieties, or producing quantities of plants identical in colour, shape or form.

CHAPTER SIX

Propagation of Conifers and Heaths

Introduction

The great majority of the conifers to be found in gardens are derived from a fairly narrow range of wild species; nevertheless, they make up a varied and mixed lot. Some are natural forms of wild plants which can be grown from seed, and which we find so attractive that we are content to grow them just as they are; others are clones propagated vegetatively from unusual plants, which have been found as sports from time to time. Some have blue or golden foliage in place of the normal green, some have finely cut fern-like shoots, or produce branchlets which hang down in a weeping form. Some are slow-growing dwarf varieties, others spread horizontally, and yet others grow rapidly to form tall spires. Many conifers can be grown from seed, some from cuttings or layers; a few are usually reproduced by grafting. There are valid reasons why gardeners choose one form of propagation in preference to another, and conifers provide good examples of how and why such choices are made. Many are tough plants, able to tolerate a certain amount of ill-treatment and can be propagated with less than expert care, and these are good plants for beginners to cut their teeth on, providing an encouraging introduction to the basic ingredients for success with cuttings of any kind.

Conifers produce seeds in cones, which are not difficult to collect and are easy to handle. Seed is an economical way to propagate large numbers of plants and virtually all the conifers in forestry plantations are produced in this way: its main drawback being the very slow growth rate of many seedling conifers during the first two or three years of their lives. However, it is a method which can be used to propagate only the naturally occurring species: these breed true from seed, producing offspring which closely resemble their parents. Although only a small proportion of the conifers grown in gardens are propagated from seed, those that are include some very beautiful and valuable specimen trees including larches, pines and firs, a few spruces and that odd-plant-out, the ginkgo.

Most of the conifers in gardens, and all the heaths, are propagated from cuttings, because, more often than not, a selected form is wanted rather than the type which grows naturally in the wild. Conifers possess remarkable versatility in their ability to change colour, shape, form, stature and leaf texture, and over the years innumerable varieties have been selected. These differ one from another in appearance and the ways they can be used in gardens, and are used very selectively to answer a particular purpose: as screens; as specimen plants; as camouflage; as a contrast in colour or shape to neighbouring plants; or as an aid to discipline in the garden, imparting neatness and order in place of that casual disarray found in less formal plants which many gardeners find so disturbing. This selective use makes it essential that each should grow up according to a precise specification so that it serves the purpose for which it was chosen. The only way to ensure this is to propagate each variety by some method of vegetative reproduction, which guarantees that the offspring grows up in the image of its parent. Many of these garden conifers produce seed regularly, and could be propagated from seed, but the result could not be predicted. Many kinds of conifers produce quantities of leafy shoots which can be used as an almost endless supply of cuttings, to provide the main means of commercial production.

However, there are some species and varieties which cannot be multiplied like this because cuttings taken from them produce no roots or produce them erratically. Some other method of vegetative reproduction must be found and most frequently the answer is to graft shoots of the selected form on to seedlings of the wild type. Plants which are produced regularly in this way include many of the ornamental forms of pines, a few firs, though very few are grown in gardens, cedars and spruces. It is also not uncommon to find dwarf or slow-growing varieties of *Chamaecyparis* grafted, inelegantly, on to seedlings of much stronger-growing wild forms. This is done to produce a saleable plant more quickly by spurring slow-moving kinds into more rapid growth. It serves the short-term interests of the seller very much better than those of the buyer and, as a rule, conifers should only be propagated by grafting when cuttings do not provide a satisfactory alternative.

Another method of vegetative reproduction, and one which is often disregarded, is to propagate by layering branches of the parents to the ground, when, after a period, many will produce roots. Most heaths and a few species of conifers, notably thuyas, multiply naturally in this way, and it is an extremely simple means of obtaining quite large plants ready-made with little trouble. On a commercial scale layering requires special techniques to produce the quantities of plants needed, and it is now seldom used by those who grow plants for sale. However, it does provide an economical and carefree way for amateur gardeners to produce a few large plants to replace losses or to provide ready-made specimens in the garden.

Seed Germination

Seeds of many kinds of conifers germinate with little difficulty, but growth during the first two years may be very slow indeed. Most of the species come true from seed, but named varieties and forms are likely to reproduce variably, and are seldom worth growing in this way. The majority are raised from seed by sowing them in nursery beds (p. 151) outdoors, or in seed pans set out in a cold frame. Most of these should be sown as soon after they are harvested as possible, and many of them produce seedlings only after they have experienced low temperatures for periods of several weeks (p. 87). A few species germinate better in a greenhouse with artificial heat (p. 56), and the results justify the extra care and costs involved. Seedling conifers should not be pricked out or potted on while they are still very small in an effort to speed up their snail-like progress, but left to grow on, at least during their first year, at their own pace.

Species that can be Sown in Nursery Beds Outdoors, or in Containers in a Cold Frame

Abies grandis

A. koreana

A. pinsapo

A. procera

Calocedrus decurrens

Cedrus atlantica

C. deodara

C. libani

Chamaecyparis lawsoniana

C. nootkatensis

C. obtusa

C. pisifera

Cryptomeria japonica

Larix decidua

L. kaempferi

Picea abies

P. brewerana

P. omorika

P. pungens

P. smithiana

Pinus griffithii

P. mugo

P. nigra

P. parviflora

P. strobus

P. sylvestris

Pseudotsuga menziesii

Sciadopitys verticillata

Sequoiadendron giganteum

Taxodium distichum

Taxus baccata

Thuya occidentalis

T. plicata

Tsuga canadensis

T. heterophylla.

Also Ginkgo biloba.

Species that Respond to Artificial Warmth (c.20°C, 68°F) and can be Sown in a Greenhouse

Araucaria araucana

Cedrus atlantica

C. deodara

C. libani

Cryptomeria japonica

Cupressus glabra

C. macrocarpa

C. sempervirens

Sequoia sempervirens

Sequoiadendron giganteum.

Propagation from Cuttings

Conifers and heaths both produce sturdy awl- or needle-shaped leaves, naturally adapted to cope with dry atmospheres, cold weather and high winds in rather exposed or inclement parts of the world. These qualities enable cuttings to function and remain alive while they develop roots, with rather less care than would be needed by more vulnerable cuttings made from the leafy shoots of deciduous shrubs. Cuttings of conifers, usually obtained from semi-mature or mature shoots, are robust and tolerant of imperfect management, an ideal introduction to the techniques of growing plants from cuttings for anyone inexperienced in plant propagation.

Selection of Material

Use of Juvenile Tissues

Plants, like animals, are juvenile when they are young but, unlike animals, it is often difficult to spot the differences that distinguish juvenile plants from adults apart from their smaller size. They are different in invisible or physiological ways, amongst which is an inability to produce flowers under any conditions. Later they become mature and will then flower as and when the time is ripe. Most annuals enjoy an extremely brief youth and it may be almost indiscernible. Tree seedlings remain juvenile for much longer, frequently for a decade or more, and some familiar garden plants like the pagoda tree, the tulip tree and the great tree magnolias may delay their first flowering long enough for it to become something of an event, not very often experienced by those who plant them.

Some species reveal their immaturity more openly by producing foliage which is distinctively different from that of mature plants, and this is particularly obvious in some of the conifers. The small plants that develop from the seeds of junipers, cypresses and thuyas scarcely look like their parents but produce much softer more feathery-looking foliage than the close-set, hard and often almost scaly structures which serve as the leaves of mature plants. Conifer seedlings which display these juvenile characteristics are, or more accurately were, sometimes known as retinospora forms of the plants concerned, and it is conventional to describe their foliage as fern-like, although no ferns exist with fronds which look anything like them. Sometimes seedlings never grow out of their juvenile plumage; occasionally shoots on mature forms of conifers revert to it; either way cuttings can be taken from them to maintain the juvenile form as a permanent feature of the variety. And, cuttings removed from juvenile shoots are usually able to produce roots more readily than those from mature ones.

Juvenile and immature are two words which are often used consecutively, synonymously and repetitively. This habit causes confusion when referring to the state of plant tissues, in which the term juvenile should be restricted to the phase, associated with seedlings, during which the plant is unable to produce flowers. Immature is a useful term to describe parts of a plant which are still growing or developing at any stage during its life span. It is sometimes worthwhile encouraging plants to produce immature shoots, because, like juvenile tissues, they often form roots more easily than mature shoots.

Effects of Size and Maturity

The shoots which are used to produce cuttings of conifers can vary in length from 3 cm (1¼ in) snippets to branchlets up to 20 cm (8 in) long. Naturally, slow-growing forms produce smaller cuttings than more robust kinds, but much of this variation can only be explained as the individual preferences of equally successful propagators. Provided that it can be accommodated under shelter and be looked after properly, the size of a cutting has much less effect on prospects of success than the variety concerned. Small cuttings need to make more growth after they have produced roots, before they can be safely planted out in the garden, and this can make a difference of a year in the time taken to produce a plant. If plenty of material is available, and only small numbers of plants are needed, it usually pays to take moderately large, rather than small, cuttings to make use of the growth the shoots made while still attached to the plant, in the months before there was any thought of using them as cuttings.

In a very similar way, heather cuttings may range in size from relatively long shoots, of about 10 cm (4 in), using almost the entire length of new growth produced during the summer, to tiny cuttings, barely 2 cm ($\frac{3}{4}$ in) long from the tips of the shoots. Both will produce roots equally successfully, but when the small cuttings start to grow their shoots will emerge close to the ground and form a bushy, stocky little plant, which some prefer to the more leggy plants produced by longer cuttings.

Types of Cuttings
As their name makes plain, cuttings are shoots (or leaves or roots) removed from their parents by cutting them off, usually with a sharp knife, or secateurs. Cuttings can be prepared by separating them from their parents by cuts made at a variety of different positions, and they are given various names depending on the way this is done.

Stem Cuttings are cut off straight across the stem, some centimetres below its tip, and, quite frequently, at the position where the current season's growth commenced.

Basal Cuttings are formed from side shoots by slicing them off at their junctions with the main stems.

Heel Cuttings are produced by pulling side shoots off, or cutting them, so that each comes away with a small sliver of the more mature tissues of the main stem.

As a rule heel cuttings are looked on with more favour than basal, and basal cuttings are expected to produce roots more readily than stem cuttings. These preferences are worth bearing in mind when plenty of material is available. If not, all three types of cuttings can be used to make up numbers, and usually the variety of plant being propagated will affect the result much more than the precise kind of cutting used.

Stage of Development
Many deciduous and broad-leaved evergreen shrubs are propagated by making cuttings from immature shoots which are still soft and flexible, but these are seldom used for conifers or heaths. Their cuttings are formed from shoots in which the woody, conducting tissues within the stem have developed to the point where they feel firm when they are pinched between finger and thumb, and noticeable resistance can be felt when the shoot is bent sideways. Some conifers, particularly junipers, reach this stage of development by late June and,

Figure 6.1 The shoots produced by conifers can be prepared as cuttings in a variety of ways. These illustrations show how the three main types of cutting – nodal, basal and heel cuttings – can be obtained from young branches of a Chamaecyparis *and a* Thuya. *Heel cuttings are usually preferred to nodal cuttings. The former are commonly used only when side shoots have grown too long to make convenient-sized heel cuttings.*

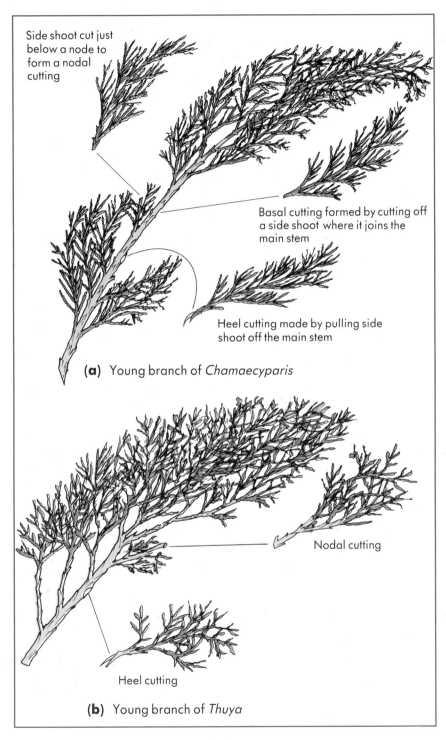

Side shoot cut just below a node to form a nodal cutting

Basal cutting formed by cutting off a side shoot where it joins the main stem

Heel cutting made by pulling side shoot off the main stem

(a) Young branch of *Chamaecyparis*

Nodal cutting

Heel cutting

(b) Young branch of *Thuya*

appropriately enough, that is the month to propagate these from cuttings. They produce roots rapidly, and once they are well-developed can be potted up and overwintered as young plants. Most conifers produce very little growth during early summer, and then grow rapidly between July and October, producing growth which is too soft for use as cuttings until it starts to mature well into the autumn. Cuttings of these can be prepared during October and November or they can be left to overwinter on their parents and removed the following March.

Heaths and heathers grow throughout the summer, and later produce their flowers from buds in the upper parts of their young shoots. A simple way to encourage them to produce plenty of well-developed sprigs which flower profusely and provide ample cuttings, is to cut them back hard every spring by clipping the plants all over with a pair of shears. Cuttings can be made from these young shoots between mid-July and mid-September. They will produce roots before the winter, but are more likely to survive if they are left in their containers, and potted up individually early the following spring.

When plants are being grown from cuttings taken during the autumn, they have to be cared for through the winter. At the least they need the protection of a sheltered cold frame, and they benefit from some form of artificial heating to protect them from the worst effects of winter cold. This is not a time for coddling with hot-house temperatures; all that is needed is just enough warmth to stop the ground freezing around their roots—by our standards, a less than tepid degree of comfort. Soil-heating cables, thermostatically controlled to prevent temperatures falling below 5°C (41°F), provide an economical and very effective solution, which can radically increase the numbers of cuttings of heaths and conifers which produce roots and survive the winter. Most cuttings of conifers taken during October and November will not develop roots until the spring, and they will be ready to pot up in early summer. Cuttings from exactly the same shoots can be made in March, after they have spent the winter on their parent plants outdoors. These can be set out in cold frames and will start to form roots within a few weeks. They need to be cared for over a shorter period than the autumn cuttings, and make no demands on cold frames or soil-heating cables, but the roots they produce will emerge a little later in the season, and the young plants will be ready to pot up a few weeks after those from cuttings taken in the autumn. This quite small delay makes a considerable difference to the size of the plant produced by the end of the first year.

Most species and varieties of *Chamaecyparis* and *Thuya*, and many other conifers, produce roots from cuttings taken in the autumn, and may be overwintered under quite demanding conditions; their tough green shoots remaining alive for long periods, if need be, until new roots are produced. Exceptions are found amongst the varieties of Leyland cypress (× *Cupresso-cyparis leylandii*). These are frequently difficult to propagate successfully from cuttings, and success is much more dependent on skill and the facilities available. Prospects are best when cuttings taken very early in February are set up, using mist propagation, and soil-warming cables to hold temperatures at about 15°C (59°F) in the cutting compost at the base of the cuttings. These plants have become so undesirably common, ubiquitously planted as hedges in town and countryside, that it may come as a surprise to find that they are not amongst the easiest of all plants to propagate.

Conifers that can be Propagated in Cold Frames or in Propagating Cases with Minimum Artificial Warmth

Calocedrus decurrens

Chamaecyparis lawsoniana 'Allumii'

  C. l. 'Columnaris'

  C. l. 'Fletcheri'

  C. l. 'Lanei'

  C. l. 'Minima'

  C. l. 'Pottenii'

  C. l. 'Stewartii'

  C. l. 'Wissellii'

C. nootkatensis 'Lutea'

  C. n. 'Pendula'

C. obtusa 'Nana Gracilis'

  C. o. 'Nana Lutea'

C. pisifera 'Boulevard'

  C. p. 'Filifera Aurea'

  C. p. 'Plumosa Aurea'

  C. p. 'Squarrosa Sulphurea'

C. thyoides 'Ericoides'

Cryptomeria japonica 'Elegans'

  C. j. 'Lobii Nana'

  C. j. 'Spiralis'

Cupressus macrocarpa 'Goldcrest'

Juniperus chinensis 'Aurea'

  J. c. 'Stricta'

J. communis 'Compressa'

  J. c. 'Depressa Aurea'

  J. c. 'Hibernica'

J. horizontalis 'Glauca'

  J. h. 'Hughes'

J. × media 'Gold Coast'

  J. × m. 'Old Gold'

  J. × m. 'Pfitzerana'

J. sabina var. tamariscifolia

J. scopulorum 'Blue Heaven'

  J. s. 'Skyrocket'

  J. s. 'Table Top Blue'

J. squamata 'Blue Star'

  J. s. 'Meyeri'

J. virginiana 'Grey Owl'

Metasequoia glyptostroboides

Picea abies 'Compacta'

  P. a. 'Nidiformis'

P. glauca var. albertiana 'Conica'

Sequoia sempervirens 'Adpressa'

Taxodium distichum

Taxus baccata 'Fastigiata'

  T. b. 'Repandens'

  T. b. 'Semperaurea'

Taxus × media 'Hicksii'

Thuya occidentalis 'Rheingold'

  T. o. 'Smaragd'

T. orientalis 'Aurea Nana'

  T. o. 'Conspicua'

T. plicata 'Atrovirens'

  T. p. 'Aurea'

  T. p. 'Stoneham Gold'

  T. p. 'Zebrina'

Thuyopsis dolobrata 'Variegata'.

NB This is only a small sample of the large number of species and cultivars of conifers which are available. Almost invariably other cultivars of the species listed could be treated in the same way.

Conifers that Benefit from Mist Propagation and Artificial Warmth

Cupressocyparis leylandii 'Leighton Green'

  C. l. 'Haggerston Grey'

  C. l. 'Naylor's Blue'

  C. l. 'Castlewellan Gold'

Cupressus macrocarpa 'Donard Gold'

  C. m. 'Goldcrest'

Heaths that can be Propagated from Cuttings

Calluna vulgaris

Daboecia cantabrica

Erica arborea

  E. australis

E. carnea

E. ciliaris

E. cinerea

E. lusitanica

E. mediterranea

E. tetralix

E. vagans

and hybrids and varieties.

Conditions that Favour Root Formation

A newly made cutting consists only of a stem, with buds and leaves; without roots it can survive for no more than a limited period, since it is roots which will supply the shoot with the water, nutrients, and hormones of one kind or another, which are essential for its growth and further development. In most cases (willows are an exception), the cutting not only has no roots, but contains no tissues, however vestigial, which were at any time destined to become roots. However, it does contain cells capable of dividing to produce new tissues. So long as shoots can be kept alive, these regenerative cells remain capable of division. When they do so, those in positions close to the base of the cutting may be reprogrammed to produce the roots on which the continued existence of the shoot depends, in place of the leaves or stems, or buds, which would have developed in the normal course of events. This is the phenomenon of totipotency in action, which enables a cell from anywhere in a plant to produce any part of a plant, or, if need be, an entirely new plant, provided it is itself capable of division. The life support system which a cutting needs in order to do this is similar to that required to enable seeds to germinate: with one important exception. Cuttings made from actively growing shoots consist of stems and green leaves; not only will the latter die if they are deprived of water, but they depend on high intensities of daylight to function effectively. We take this for granted with plants that are growing naturally out of doors, but some of the biggest problems met with when trying to persuade cuttings to produce roots are due to their mutually incompatible requirements for high light intensities, which tend to produce dry conditions, and a humid, virtually saturated atmosphere around the leaves.

The major physical conditions that affect the behaviour and responses of cuttings are:

Water. This must be freely and consistently available, not only in the compost at the base of the shoots where it is taken up to supply the leaves, but also in the atmosphere around the leaves as water vapour, to reduce the rate at which they transpire.

Oxygen. Supports the vital processes of respiration. Its presence in adequate amounts is essential around the base of the stem, in the region where new roots will be produced. This may be 4 or 5 cm ($1\frac{1}{2}$–2 in) below the surface of the compost.

Light. Photosynthetic and other processes depend on high light intensities to function effectively. Photosynthesis is the main means by which plants obtain energy, which provides an essential fuel for all the processes involved in growth and development.

Temperature. The initiation and development of new roots are processes which involve chemical and enzyme reactions and these proceed most rapidly under warm conditions. The temperatures which promote growth most effectively, and economically, usually lie between 12° and 22°C (54–72°F). Lower temperatures prevent or prolong root development; higher ones may inhibit it.

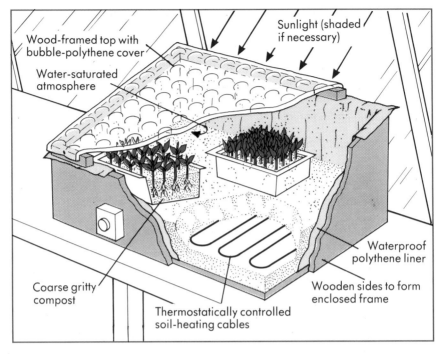

Figure 6.2   Most cuttings root best in a humid enclosed atmosphere, with warmth and light.

Cuttings respond to the physical conditions of their environment in the ways outlined above: their readiness to survive and produce roots is also affected by their well-being, or physiological state, including the balance of hormones in their tissues, and their nutritional condition.

Effects of Hormones on Root Formation

All plants depend on chemical messengers called hormones to co-ordinate the ways they develop and balance the growth of different, apparently independent, parts like the roots and shoots. Hormones move from one part of the plant to another, providing an information system which ensures that the plant as a whole responds in a coherent and appropriate way to the situation in which it is growing. The precise ways in which this system works may still be obscure, but enough is understood to allow our knowledge to be put to some practical use.

Hormones influence the ways that cells divide, and the nature of the tissues they give rise to, and are crucially important in directing and controlling the processes by which cuttings recognise the absence and need for roots, and set about the business of producing new ones. Observations suggest that roots are produced in locations where high levels of particular kinds of hormones, known as auxins, occur naturally. These control the development of cells after

74

division in such a way that they produce root initials, or embryonic roots: and it is these which grow into the fully formed emergent roots on which the survival of cuttings depends. Distinguishing between the production of root initials, and the outgrowth of the root itself, might appear to be as punctilious as splitting hairs, because for almost all practical purposes they are part of a continuous process—but for one thing. Paradoxically, high concentrations of auxins appear to encourage the production of root initials, but they also prevent their further development, and low levels are needed before the roots are able to develop and grow properly.

Synthetic chemicals, which reproduce the effects of these naturally occurring auxins, are available and are used by gardeners to encourage cuttings to produce roots. Sold as 'Rooting Hormones' or 'Rooting Compounds', they are prepared from several different, but structurally related, chemicals, which have similar effects, but which vary in their effectiveness and the ways they are used. Some are formulated as powders, some as liquids: the concentrations may vary to take account of the needs of different plants, or this may be achieved by using chemicals of greater or lesser activity. It is essential to use all of them exactly as recommended by their manufacturers, and to avoid the common inclination of all gardeners to add a bit more for luck—and then just a bit more to make sure. They should be applied only to the base of the cutting, since that is where roots are to be produced, and they should be applied sparingly, bearing in mind that all that is required is an initial boost to supplement the effects of the plant's own naturally occurring auxins. An over-generous helping may prevent or deform the development of the roots.

The chemicals which are used as rooting hormones are not reliably stable, and can deteriorate quite rapidly, especially at high temperatures or in bright sunlight. The potting shed window sill, or a shelf in the greenhouse, are bad places to keep them, but they can be stored, dark and chilly, in a refrigerator for at least twelve months without becoming ineffective.

The auxins, which encourage cuttings to produce roots, have many other parts to play in plants. One is concerned with the restoration and repair of damage, and a natural consequence of a gashed stem or a crushed root is an increase in auxin concentration in adjacent cells. This leads to renewed cell division, and the formation of new tissues which seal off and replace those which were destroyed. Gardeners sometimes use this response to wounding by removing one or two thin slivers of bark from the side of a cutting for 2 or 3 cm ($\frac{3}{4}$–1$\frac{1}{4}$ in) above its base. They hope that the wound produced will stimulate the cutting to produce natural auxins, referred to as 'wound hormones' and stimulate root formation. The method is usually reserved for cuttings which have grown woody with maturity, or which naturally produce very strong hard stems. Sometimes the conducting tissues within these form a complete ring, encircling the cambial cells from which roots are produced, and making a barrier through which the roots cannot escape. A sharp knife will slice through these woody tissues, providing a way out for roots which might otherwise be trapped. Frequently it is hard to say whether any improvements which follow are due to the effects of wound hormones, or the removal of a mechanical barrier. Either way, wounding a cutting is a simple means of improving the performance of cuttings which are reluctant to root or which do so very sparingly.

Effects of Nutrition on Root Formation

The nutritional condition of plants, whether they are well fed or starved half-way to death, can affect the way that cuttings taken from them produce roots. Plants in a verdant, lush condition which are responding to high levels of nitrogen in the soil may produce numbers of very healthy cuttings, but these are unlikely to form roots readily. Starved plants, making slow growth, produce small numbers of cuttings, which may be so hard and stunted that it can be difficult to induce them to make growth of any kind. The ideal is to take cuttings from plants which are growing steadily under reasonably fertile, nutritionally balanced conditions and to select the shoots used as cuttings by rejecting any which are unusually large, and healthy looking, as well as nutritionally deprived, wizened and spindly twigs.

Traditionally it has seldom been considered necessary to add nutrients to the compost used for cuttings, the assumption being that since cuttings are fairly substantial bits of plant, they can draw on their own internal resources during the period when they are initiating roots. Once new roots are formed, and the shoots start to grow again, there is no question that the rooted cuttings benefit from being fed (p. 82).

Recent observations have shown, however, that contrary to tradition the production and growth of roots can be improved by adding fertilisers to cutting composts from the very beginning. This can cause problems. Well-made cutting composts, containing plenty of grit and open pore space and very low levels of nutrients, are not hospitable places for the germination of the spores of mosses and liverworts. As a result they are only slowly invaded by these pests which can cover the surface in a dense green mat, clogging pore spaces, and reducing the diffusion of oxygen into the compost. The addition of nutrients to the compost immediately makes it a much more attractive home for the development of these undesirables, and more often than not its surface rapidly becomes covered in a green film. This is unlikely to matter when the cuttings produce roots rapidly, and grow away strongly. It becomes much more serious when cuttings are slow to produce roots and then the beneficial effects of adding nutrients to the compost can be outweighed by the unwelcome presence of the mosses. On balance it is advisable not to add nutrients to cutting composts, but to wait until roots are formed and the cuttings have started to grow, and then make very sure that the developing cuttings do not suffer at any time from starvation.

Composts for Cuttings

A very wide variety of different materials can be used to provide the substrates in which to set cuttings while they produce roots, and there is little evidence that any particular one possesses the magic touch, which makes it the indispensable mainstay on which successful results depend. Nevertheless, success comes easier if composts are prepared which can be managed, without unnecessary care and attention, and which spontaneously provide conditions which enable cuttings to survive and for roots to form and develop with as little stress as possible. The compost must be sufficiently dense to support cuttings mechanically so that they do not fall over; it must be open enough to allow air to percolate through it freely; and yet be able to hold as much water as possible for as long as possible.

Grits of one kind or another provide a good starting point for a cutting compost. They are available. They may be acidic or basic so that it is possible to choose one that suits the needs of plants like rhododendrons or the summer-flowering heathers which only thrive in acid composts, and a different kind for lavenders and rosemaries which have a preference for basic conditions. They possess many of the properties needed to provide the conditions that cuttings need to produce roots successfully. The graded sands used by builders are much less satisfactory, as is silver sand. Silver sand—the words have a hint of magic—has been recommended parrot fashion, and repeatedly, as an aid to propagation whenever the topic has been raised, and has become firmly set in gardening lore. But this constant reiteration of its virtues need not persuade us that it is the best material we can find for the purpose. The individual grains of all these sands are small, and closely similar in size, so that when they are watered they settle into a closely packed mass; each one fitting into place like pieces in a jigsaw puzzle with very little space between it and its neighbours. This provides the qualities needed to make good quality cement, which depends on gluing separate grains together to form a solid mass. This is not a property which makes a good cutting compost. The small spaces between the grains of sand fill with water as soon as they are watered, and this is held by surface tension, until the whole mass dries out, uniformly and rapidly as the water is replaced by air. Consequently the compost provides only two alternatives—it is either extremely dry or extremely wet, and the change from one condition to the other is very rapid. Grits, made up of particles varying in size from about 8 mm ($\frac{1}{3}$ in) down to dust, pack together in a haphazard fashion, some particles nestling closely together, others held apart by their sharp corners, or by smaller grains lodged between them. The smaller spaces within the mass fill with water held between adjacent grains by capillary attraction; the larger spaces become tiny air-pockets. At any particular time there tends to be a mixture of air and water held in the various-sized spaces present in the bulk of the compost.

The small particles of broken rock which make up most grits absorb very little water, and composts containing nothing else dry out rapidly. The usual answer to this problem is to add a proportion of peat, which absorbs water like a sponge and greatly increases the water-holding capacity of the compost. Peat is used very successfully in many nurseries, but for amateur gardeners may bring problems, which can cause poor results or failures.

Additions of peat inevitably counteract the good qualities of the grit by clogging spaces between the particles with fibres and peat dust. When the compost is overwatered, these compact into a mat through which air diffuses slowly, and this is accentuated as the peat deteriorates, and its surface becomes a nursery bed for the development of mosses and liverworts. If, aware of these dangers, the gardener errs on the stingy side when watering, the mixture dries out and becomes extremely difficult to rewet thoroughly and evenly. Peat's intolerance, both to overwatering or underwatering, is well known. However, other water-absorbing materials which can be substituted for peat are available and do not produce its problems. These are much easier to manage because they can be watered freely with little or no danger of waterlogging, and, when they dry out, can be easily and uniformly saturated with water again.

The most easily managed alternatives to peat are porous, gritty materials of one kind or another (p. 56). These include calcined clays (bentonite, montmorillonite, etc.), and materials of volcanic origin including perlite and vermiculite. Any of these can be added at between 30 and 50 per cent of the total bulk to improve the water-holding properties of grit-based composts, without reducing the desirable properties of the grit. They drain freely and rapidly, so that overwatering does not result in poor aeration, and take up water again easily after they dry out: it can be taken for granted that sooner or later they will dry out. A good idea, to tell quickly and easily whether composts are becoming dry, and need to be watered, is to use a material which changes in colour, or at least in intensity of tone as it becomes dry.

Setting Out Cuttings

The cutting composts described above combine properties which ensure that the part of the cutting where roots will be produced is always in contact with a supply of oxygen, and at the same time is not too readily deprived of water. These needs for water and oxygen should be borne in mind when cuttings are set out; they should be inserted no deeper than is necessary to hold them upright—tiny heather cuttings perhaps with their bases only 1 cm ($\frac{1}{2}$ in) below the surface; larger cuttings of conifers pushed down no more than 3 or 4 cm ($1\frac{1}{4}$–$1\frac{1}{2}$ in). Cuttings will support each other mechanically if closely packed, and this also helps to maintain a close, humid atmosphere around their leaves. A 9 cm ($3\frac{1}{2}$ in) square pot should be looked on as a suitable container for something like 30 heather cuttings or 12 sprigs of one of the larger cultivars of Lawson's cypress.

Caring for Cuttings: Equipment and Facilities

Once in place cuttings should be left undisturbed. It will take them at least a few weeks, perhaps several months, to produce roots; but, human nature being what it is, the temptation to pull up a few cuttings to see what they are up to must be resisted. These disturbances disrupt contact between the ends of the cuttings and the film of water held by the particles in the compost, and can reduce the chances that roots will be produced eventually. As a rule, a glance at their foliage will reveal how cuttings are progressing. Newly made cuttings always appear spick and span and glowingly healthy, they are a credit to any gardener and look as though they couldn't possibly fail; then there comes a period when they lose their gloss, and may look a little drab, and doubts about the outcome succeed optimism. But once roots have formed there will be an improvement in 'tone' as the foliage regains the glossy sheen typical of the leaves of healthy plants and, even more significantly, the tips of the cuttings begin to grow again and the fresh green colour of young emergent leaves becomes very noticeable. It is unusual for cuttings to make any new growth until they have produced roots, and once this is noticed a gentle tug which meets resistance will confirm their presence; it is not necessary to haul the whole cutting right out of the compost. As long as cuttings remain green, and their stems stay healthy and undecayed, there is hope and a chance that they will produce roots. Conifers, with their resistant, tolerant and enduring foliage, may occasionally take 18 months or more to do so.

The trick depends on finding ways to keep these dismembered shoots alive and functioning as normally as possible without causing them needless stress until they produce roots. Stress, caused by, amongst other things, lack of water, adversely high temperatures, or shortage of oxygen, reduces the ability of cuttings to function adequately, increases their susceptibility to pathogens and, overall, increases risks of failure.

What is needed is the plant's equivalent to an incubator for a premature baby, a well-lit, enclosed container in which to put the cuttings, and keep them warm and moist. Continuous high humidity in the air around the shoots is the first essential otherwise they will lose water, cease to function and die. In practice this makes it very difficult to provide any ventilation, unless expensive automatic watering systems are used to provide a constant mist, or fog, of water vapour. Sunlight is the most economical and effective source of warmth and bright light, but, without ventilation, produces the same conditions as an overheated turkish bath which no cuttings will survive for long. Diffused sunlight is easier to manage and unlikely to damage young cuttings. This is why, when gardeners use sun-frames, the glass is painted with a slick of whitewash during the summer.

The simplest way to make an incubator is to enclose the cuttings, set up in a container, in a plastic bag, and tie up the open end to prevent water vapour escaping. This can be used temporarily to bring cuttings home from a friend's garden, or keep them safe for a few days, provided the ensemble is kept out of direct sunlight. The plastic bag enclosing a potful of cuttings is frequently put forward as a useful way for amateur gardeners to produce plants from cuttings, but unless the number needed is very small indeed, it is too fiddly and unnecessarily awkward to bother with.

Sheets of glass can be used to enclose cuttings but plastics like polythene, polystyrene or acrylic sheet are lighter, much easier to work with and safer. These can be used to construct frames of any size, depending on what is needed. Perhaps the ideal solution when small to moderate numbers of cuttings are being produced are the moulded polystyrene propagators specially made for the purpose. They lend themselves to flexible management and are convenient to use and to accommodate, whether pushed beneath a bench in a greenhouse, set out in a shaded cold frame or put in a cool place in a corner of the garden.

More often than not cuttings are taken in batches. An afternoon is spent going around the garden taking cuttings from a dozen or more different shrubs and plants; an opportunity may be offered to raid a friend's garden, and should be accepted appreciatively and acted on eagerly. After all, not to do so would suggest that there was little worth having. Once the collections have been made there is a temptation to set to, fill the base of a propagator with the chosen cutting compost, and line the batches of cuttings out side by side in neat rows; then sit back to admire a job that looks well done. A few weeks later, the shoots

Figure 6.3 The simplest way to root cuttings of many kinds is in a closed container, such as a propagator with a moulded polystyrene top. Different kinds of cuttings must always be set up separately in their own containers. Cuttings will usually produce roots more rapidly and more successfully when given artificial warmth.

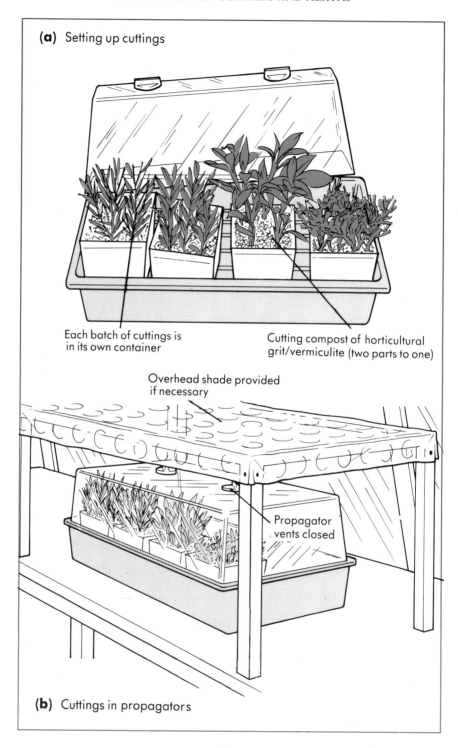

(a) Setting up cuttings

Each batch of cuttings is
in its own container

Cutting compost of horticultural
grit/vermiculite (two parts to one)

Overhead shade provided
if necessary

Propagator
vents closed

(b) Cuttings in propagators

of some batches of cuttings start to grow away, showing that they have produced roots, others remain unchanged with no suggestion that they have produced, are about to produce, or ever will produce anything like a root. This creates a dilemma, and teaches a useful lesson. The dilemma is caused by doubts about what should be done when some cuttings in a container are ready to be fed and moved on, and others in the same container are nowhere near that stage, and should be left where they are. The lesson is never to put cuttings of two different kinds of plant together in such a way that they cannot be managed individually. There is no objection at all to a dozen or more different sorts sharing the same propagator or framespace provided each lot is set up independently in its own separate container.

Mist Propagation

Mist propagation units are widely used by commercial growers, and similar equipment, marketed for amateurs, is often looked on as the ultimate indulgence and one which holds the golden key to success. Mist can be an exceedingly useful facility and some kinds of plants are difficult to propagate without it. It is not necessary for the great majority of cuttings, which will produce roots quite satisfactorily enclosed in saturated atmospheres beneath sheets of glass or plastic film. A mist propagation unit, including the costs of its installation, is a fairly expensive piece of equipment in itself, but has very limited value unless set up in combination with soil-heating cables, which can be thermostatically controlled to maintain temperatures of 10° to 20°C (50–68°F) in the cutting compost around the base of the cuttings. The total costs of setting up and running such a propagation system can be quite considerable; should always be thought about carefully; and can be justified only when a considerable number of cuttings are to be produced from plants which do not respond to simpler methods.

The key to the success of mist propagation lies in the fine spray of water droplets which saturates the air around the cuttings. These supply the answer to the conundrum of simultaneously maintaining high light intensities and high atmospheric humidities, and cuttings receiving the benefits of mist do not have to be kept in a shaded, enclosed space, but are simply set out on a bench with the protection of a greenhouse to shelter them from cold and wind. The light intensity, from sunlight, is kept as high as possible with no more than a little shading to protect the cuttings from its force; and it enables cuttings to continue to function at almost normal light levels without shrivelling up and dying.

Undeniable though these advantages are, they bring dangers with them. The great limitation of mist propagation is that it is not a fail-safe technique. When anything goes wrong disasters follow. If the unit fails to operate, the unprotected cuttings dry out very rapidly; alternatively, if malfunctions cause the unit to switch on more frequently than it should, the compost becomes saturated with water, and the basal ends of the cuttings are deprived of oxygen. If regular supervision is not possible, more passive systems, based on enclosed saturated atmospheres, are not only much cheaper to set up and operate but provide few opportunities for things to go wrong.

Nutrition of Rooted Cuttings

There may be good reasons for disregarding advice to provide cuttings with nutrients while they are in the process of forming roots. There is no doubt at all that, once they have produced roots, it is vitally important to feed them, or they will make only limited growth. The most satisfactory way to do this is with a high potash, low nitrogen, soluble feed like Phostrogen or a liquid feed of the kind used with the growbags in which tomato plants are grown. This is one instance where the quantities used should greatly exceed the manufacturers' recommendations. Most cutting composts contain no appreciable nutrients of any sort, and the first feed applied must be strong enough to remedy a more or less total deficiency. It will do so only if it is applied at three to five times the concentration suggested for normal circumstances, when it is being used to top up falling, but not virtually non-existent, levels of fertility. Subsequently, rooted cuttings should receive liquid feeds at recommended strengths every fortnight throughout the growing season, until the cuttings are potted up.

Potting Up and Growing On Rooted Cuttings

Cuttings which have produced roots develop rapidly to become self-sufficient but they do require a short weaning period before being exposed to the rigours of a completely unprotected environment. Cuttings which have produced roots successfully should be removed from the enclosed propagator or frame in which they were first put and moved to another, where a little ventilation is provided day by day. This might be a propagator with the vents permanently open, or a frame in which the lights are partially opened for several hours daily. After a week or ten days of this weaning process the cuttings should be sufficiently adjusted and established to put up with the less protected conditions of a greenhouse bench, a more or less continuously opened cold frame, or even a sheltered place out of doors. Provided they are kept watered and not allowed to become starved, they will continue to grow steadily in the cutting compost in their original containers, and are in no danger. Sooner or later they will have to be potted up individually, but there is no need to rush into this. Potting up rooted cuttings causes far more losses than need be, and there are circumstances where more plants are lost at this stage than during the apparently more risky processes, when cuttings are dependent on producing roots to survive.

Cuttings, junipers for example, which produce roots early—by July or into August—make much faster growth and develop into larger plants if they are potted up as soon as their roots are well enough developed to benefit from the move. Other conifers, whether taken as cuttings in the autumn or the spring, respond to being potted up as soon as possible in the summer and their ability to grow through August and September, and well into October, ensures that they make good use of the opportunity. Heather cuttings produce vulnerable, fine hair-like roots and delay doing so until August or September or even later. They will make little or no growth after being potted up. The compost will lie wet and soggy around their barely developed roots all winter, and most will have given up the struggle by the time spring comes. It is far safer never to attempt to move these on until the pale tips of the developing shoots the following spring show that they have begun to grow again after the winter.

The first winter is a testing season for a recently formed rooted cutting and the time it is potted up can easily make the difference between its survival and death. Young, recently potted cuttings can be exposed to risk, stress and probable early death in a number of ways, and these are summarised below:

- Potting up too soon, when many cuttings have only a few barely developed roots and some have none at all. It is worth making sure that most of the cuttings in a batch have well-developed roots before starting to pot them up.
- Moving cuttings into containers which are too large. The smallest container which will accommodate the roots, and allow for a little growth, is the safest. Inevitably this means that they will soon have to be moved on, and this should be done when the roots have completely encircled the ball of compost within their pot, but before they become densely matted and entangled.
- Potting up late in the growing season. This applies particularly to plants with very fine roots like heathers. As a simple rule of thumb, 1 September provides a sensible deadline after which no cuttings should be potted up. The plants will overwinter more safely in their cutting compost than in a richer potting compost in which they have not become well-established. It is also easier to find shelter for one small square pot, than for 20 small plants each in its own container.
- Potting up and exposing to more rigorous conditions simultaneously. Some protection should always be maintained after cuttings are newly potted to help them become established. And preferably, newly potted cuttings should be returned to the same environment that they were in before being potted.
- Failing to ensure that the potting compost is fully saturated with water after cuttings are potted up. This can be particularly lethal with peat-based composts in which the surface layers are easily moistened, but lower layers can be bone dry.

Growing On in Nursery Beds

Cuttings which have been potted up, and are ready for their next move, are known as liners; quite literally because this was the stage, and still is in places, where they were lined out in rows in the field to grow on and produce open-ground saleable plants. These days most of them are moved on into a larger pot, and it has become easy to think of this as the natural way to do things. But before automatically buying a load of expensive pots, and large quantities of potting compost to carry these liners on for another year or two, it is well worth thinking about the alternatives. It is not at all easy to look after plants growing in containers. They depend on time and regular attention, which is seldom available, and unless capably watered, which is one of the most demanding and difficult of all gardening skills, and fed when they need to be, they will fail to prosper and may develop inadequately. It is usually easier to set young plants out in a well-prepared nursery bed (p. 31) outdoors where they will continue to develop with fewer demands on care and attention.

Shelter from cold winds makes a great difference to the rate at which the plants will grow, and even in a well-enclosed garden it is often worthwhile protecting the young plants with a shelter of hurdles, when appearance counts, or wind-break netting when it doesn't. A trellis with ivy growing up it occupies virtually no space, and provides a decorative as well as an effective way of isolating a small space within a garden, which can be used as a little nursery. The ground should not dry out, and a supply of water should be available, or the ground kept very well mulched between the plants. It is usually better if the plants are grown in narrow raised beds. These always provide good drainage; they simplify maintenance; and they are very adaptable. It is also well worthwhile making an effort to improve the texture and fertility of the soil in the nursery beds by adding old potting compost, or grit or sand to tacky clays, or plenty of humus and compost to hungry gravelly or sandy soils.

The plants grown in nursery beds cannot so easily be lifted and transplanted into the garden at any time of the year, like plants in containers. Generally speaking this is better done between late September and early March; though many plants are very tolerant of being transplanted rapidly, firmly and confidently even when in full leaf and active growth. But, they will most probably grow very much better than home-made container-grown plants, and, perhaps even more important, will re-establish themselves when they are planted out with an ease which will surprise anyone whose previous experience has been limited to container-grown plants, raised, perhaps, in New Zealand, flown half-way around the world, dunked into a pot with a dollop of peaty compost during a transit stop at a wholesalers, and finally sold off in a garden centre.

Division

Layering Conifers and Heaths

Thuyas can be layered, using similar methods to those described later for shrubs (p. 143), and this is a straightforward way of making a few large plants from existing specimens to plant out elsewhere in the garden.

Heaths of all kinds frequently layer themselves as their spreading, procumbent branches lie in contact with the ground. This natural tendency can be encouraged by shovelling over and raking into established plants a mixture of peat and grit, and in parts almost burying their shoots. Even mature and very woody stems will produce roots, and quite large and well-developed branch systems can often be separated, ready-rooted, to be planted elsewhere. This may be a tempting prospect for anyone in need of an instant garden, but the most successful layers are usually the smaller ones. Young shoots with well-developed roots at their base can be lined out in a nursery bed and will grow on rapidly to form well-shaped, bushy plants. More mature systems often establish with difficulty and suffer setbacks in the process from which they never really recover.

CHAPTER SEVEN

Propagation of Alpines

Introduction

Plants which gardeners group together in a rather arbitrary way because they grow naturally in mountainous regions are referred to as alpines. They are found in many, and diverse, parts of the world, and share little in common apart from their ability to survive under extremely rigorous, exposed conditions. These may subject them to very severe stress from high winds, long periods of snow cover, or sub-zero temperatures, extreme variations in temperature between day and night, or prolonged subjection to drought or persistent rain or mist. In addition to these stalwarts there are a great many others which might be called 'honorary' alpines. These are plants which come from temperate regions of no special severity, but are low-growing species which fit comfortably into the gardener's concept of what a plant that could be grown on a rock garden should look like. These have no claim to be alpines in any natural sense, but are not out of place filed as rockery plants.

Alpines can survive the rigorous climatic conditions of their homelands only because they have been able to develop a life cycle, or a physical form, which complements the main features of the environment; mitigating the effects of its severities, and taking advantage of periods when conditions are more benign. It is a characteristic of plants found in such situations that almost every aspect of the ways they adapt is devoted to surviving climatic hazards, and the plants are very vulnerable to competition from their neighbours when grown under less extreme conditions.

High altitudes, in common with high latitudes, are parts of the world where favourable conditions for growth are likely to be short-lived. Plants may be able to flower and produce seed successfully, but by the time the seed matures, the imminent onset of winter provides insufficient time for seedlings to emerge, establish themselves and accumulate the food reserves they would need to survive the long period of inclement weather ahead of them. They have better chances of survival if their seeds lie in the soil until the following spring, and then germinate when milder weather returns, and the whole, if brief, summer is before them. Many alpines achieve this by producing seeds which are incapable of producing seedlings at, or soon after, the time they are shed, but can do so once they have experienced a period of cold weather, at temperatures close to freezing point, for periods of several weeks or even months.

Seeds which do not germinate more or less rapidly after being sown are customarily referred to as 'dormant'. The fact that they do not germinate is described as 'failing' to germinate, and conditions (like the low temperatures described above) which produce changes leading to germination, are said to have 'broken' dormancy. Although the words used are all well established they convey a misleading view of the ways that seeds contribute to plant survival by sensing and responding to their environment. The word dormant with its indication of sleep fails to express the fact that the seed lying in the soil constantly and actively senses what is going on around it, and responds to its situation in positive and often very subtle ways: a situation to which the word latent might more aptly be applied. The inference that seeds which do not germinate have failed in some way springs from a fundamental misunderstanding of their natural function, and pays no attention to the strategic significances of this event in the plant's life cycle. Under natural conditions seeds are presented continuously with the option to germinate or remain as they are. Failure could be an appropriate description only when the option chosen is one which exposes the emergent seedling to the greater risks of danger, and it is only under the extraordinary conditions of cultivation by man that germination becomes unequivocally desirable. Breaking dormancy suggests an action which is violent and irreversible, whereas the processes which condition seeds and change their responses to subsequent events are subtle, and varied, and often easily reversible.

Just as seeds must possess appropriate germination strategies to survive, and just as these must be understood by gardeners who want to grow alpines from seed, so the vegetative growth of the plant is modified by, and responds to, its environment. Alpine plants are frequently extremely compact, forming dense mounds of closely ranked foliage on multiple crowns that hug the ground. This reduces the surface area of the plant exposed to the desiccating effects of wind or intense cold; sometimes reinforced, as in some of the saxifrages, by the possession of leaves whose form and anatomy is also adapted to resist drought and cold. This compact form is produced when stems scarcely elongate at all, with the result that the leaves remain in closely packed whorls. A rosette removed from such a plant is directly comparable to a side shoot taken from a shrub and can serve the gardener's purpose as a cutting in just the same way.

Many alpines depend on a totally different strategy for their survival; these are capable, when the need arises, of growing exceptionally long stems to enable them to cope with the shifting, unstable conditions, and many have a very well-developed facility for producing roots along these stems. This not only establishes them in newly deposited layers of debris, but provides gardeners with a very simple means of propagation.

Seed Germination

Alpines that are Easy to Propagate

A great many of the plants grown in gardens as alpines germinate very easily, under the same conditions as those used to produce seedlings of hardy annuals (p. 56). All will germinate more rapidly when warm than when cold, and temperatures between 12° and 20°C (54–68°F) provide a favourable range, which ensures rapid germination and vigorous seedling development.

Plants Grown as Alpines that Germinate Easily

These can be sown in a greenhouse or cold frame and treated as though they were hardy annuals.

Acaena*	Diascia	Parochetus
Achillea	Draba aizoides	Penstemon heterophyllus
Aethionema armenum	Dryas	P. roezlii
Alyssum	Edraianthus	Potentilla aurea
Aquilegia alpina	Erigeron aurantiacus	P. fragiformis
A. canadensis	E. mucronatus	Scabiosa alpina
Arabis	Erinus	Scutellaria
Armeria	Erysimum	Sedum
Aster alpinus	Gentiana cruciata	Sempervivum
Aubrieta	G. kurroo	Silene alpestris
Campanula barbata	G. parryi	S. elizabethae
C. carpatica	G. pneumonanthe	S. schafta
C. garganica	Geranium	Sisyrinchium
C. rotundifolia	Helianthemum lunulatum	angustifolium
Carlina acaulis	H. serpyllifolium	S. bermudianum
Cerastium	Iberis gibraltarica	Trifolium uniflorum
Dianthus alpinus	Leontopodium	Veronica
D. crinitus	Linaria alpina	Viola biflora
D. deltoides	Linum alpinum	V. cucullata
D. haematocalyx	Ourisia	V. labradorica
D. monspelliensis	Papaver alpinum	V. odorata
D. neglectus	P. pyrenaicum	V. septentrionalis
D. superbus	P. rheticum	Wulfenia.

* References here and elsewhere to a generic name only indicate that many species within the genus, but not all, will germinate under the conditions indicated.

Alpines that Require Special Treatment

Numerous alpines produce seeds which are not so easy to germinate, but will produce seedlings only after they have experienced fairly lengthy periods at low temperatures. During this time, covert changes, known as conditioning effects, play their part in three separate and successive phases:

1. *Preliminary phase.* This starts as and when the seeds are shed by their

parents. Seeds are able to imbibe water and become fully hydrated, but do not germinate*, however favourable their situation may seem to be.

2. Conditioning phase. The seed responds to some feature of its environment (e.g. low temperatures) without changing visibly, but in ways which may enable it to germinate* later. Conditioning is usually a lengthy process requiring periods of four to ten weeks to be effective.

3. Concluding phase. The radicle and plumule emerge from the seed to produce a seedling.

* Conditioning treatments can also act in a negative way by leading to physiological changes which render seeds, which are capable of germination, less likely to germinate.

Conditioning effects only happen when seeds are fully imbibed with water. When seeds are stored dry in a refrigerator, or even in a deep freeze, their capacity to germinate will not be changed in spite of frequent suggestions that cold storage of this kind is effective. Sometimes gardeners refer to this process as 'freezing' seeds, especially where alpines are concerned; the implication being that temperatures at or below freezing point are the most effective. This is not so: temperatures just above freezing point around 2°–4°C (36–39°F) are more likely to produce the changes required.

It is customary to sow alpine seeds early in the year—January is usually suggested, and December is better—in containers placed in a cold frame. Normally the weather then provides the seeds with the cold exposure they need, so that they germinate when temperatures start to rise later. The seeds should be sown as described earlier (p. 56) and their containers set out in the frame, and thoroughly watered. They can be left there until seedlings appear as the weather becomes warmer during April and May, but they will grow quicker if brought into a greenhouse at the beginning of April, and encouraged to grow on more rapidly with enough warmth to ensure that temperatures do not fall below about 10°C (50°F) at night. As a rule alpines do not benefit from artificial warmth at any time, but this is one occasion in their lives when they appreciate and can make use of a little extra comfort. Tiny seedlings of alpines are very vulnerable to pests, and it is worthwhile using a little extra warmth to shorten the time when they are very small and most at risk.

This simple, practical way to germinate alpine seeds that need a cold conditioning treatment can work well. However, there may be snags; the seeds and young seedlings are exposed to the predatory activities of slugs, mice and

Figure 7.1 Many kinds of plants produce seeds that will germinate only after they have received a conditioning treatment. The illustration shows how conditioning treatments can be given to seeds which respond to low temperatures, using a refrigerator. Some seeds require high temperatures during their conditioning phase; these can be processed in the same way, using an airing cupboard, or an electrically heated propagator instead of a refrigerator.

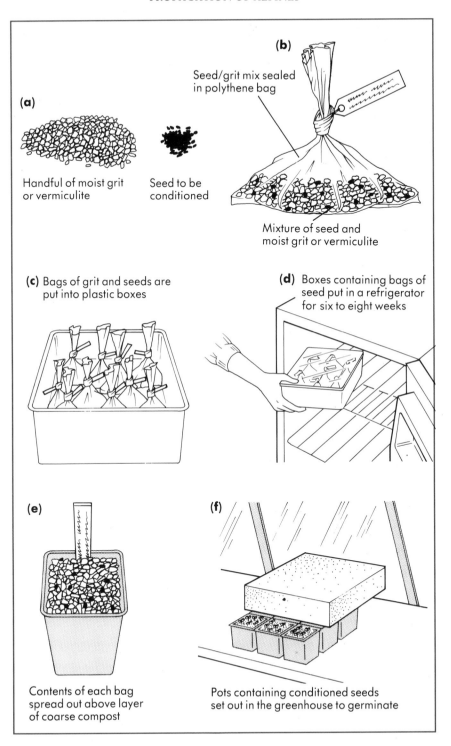

(a)

Handful of moist grit or vermiculite

Seed to be conditioned

(b)

Seed/grit mix sealed in polythene bag

Mixture of seed and moist grit or vermiculite

(c) Bags of grit and seeds are put into plastic boxes

(d) Boxes containing bags of seed put in a refrigerator for six to eight weeks

(e)

Contents of each bag spread out above layer of coarse compost

(f)

Pots containing conditioned seeds set out in the greenhouse to germinate woodlice, and to fungal pathogens while they are in the cold frame. They are also dependent on the weather and must endure whatever it brings. There are occasional years when temperatures fail to remain low enough for long enough during January and February to condition some seeds effectively; in other years violent fluctuations, or very late frosts, kill young seedlings. The method also stringently restricts the time of year when seeds can be sown successfully; even mid-February is too late in most years to provide the lengthy cold periods needed.

These uncertainties and restrictions can be entirely avoided by conditioning seeds artificially, using a refrigerator to provide them with the low temperatures they need. The seeds are mixed with a convenient quantity of moist grit or vermiculite, about a handful is usually enough, in a strong polythene freezer bag, and the top closed by tying it in a knot. The bags are then packed into plastic sandwich boxes, or food containers, and put into an ordinary refrigerator, for as long as is needed for the conditioning treatment to be effective. Refrigerators usually run at between 2 and 5°C (36–41°F) and provide temperatures close to those at which conditioning effects work best. The bags should be left in the refrigerator for six to eight weeks, and can then be removed, opened up and the contents distributed over the top of a small pot, above a layer of potting compost. If they are then watered heavily the top dressing containing the seeds will settle as it drains to provide a firm seed bed to support the seedlings. During the concluding phase of the process the pots containing the germinating seeds can be placed in a propagator or on the bench of a greenhouse, preferably with a little warmth to encourage the development of the seedlings after they emerge.

Alpines that Germinate Only After Conditioning Treatment

Aethionema grandiflorum	Erodium	Primula auricula
A. pulchellum	Gentiana acaulis	P. clusiana
Androsace	G. asclepiadea	P. farinosa
Anemone magellanica	G. lutea	P. glaucescens
A. multifida	G. septemfida	P. minima
A. ranunculoides	G. verna	P. spectabilis
Aquilegia atrata	Haberlea	Pulsatilla alpina
A. einseleana	Hypericum coris	P. halliana
A. chrysantha	H. olympicum	P. vulgaris
A. discolor	H. rhodopeum	Ramonda
A. scopulorum	Lewisia	Saxifraga
Corydalis	Linum arboreum	Soldanella alpina
Cyananthus	Phyteuma	S. montana.

Propagation from Cuttings

Propagation of Alpines from Semi-mature Stem Cuttings

Alpines amongst high mountain screes survive and make their living in an unstable environment. Stones and debris may bury them from above in avalanches and landslips; they may be washed from their rootholds by torrents from melting snow and deposited amongst tailings of clay and shattered rocks. Their shoots may be bruised, broken or flattened sideways, be overlaid by snow and left pressed to the ground. Survival depends on adapting to these hazards, and even making use of them, e.g. partially broken shoots or flattened stems produce roots and layer themselves naturally. These processes are used by the gardener in a more artificial context, during which cuttings are made, which also produce roots and establish themselves as new plants.

Throughout the summer, shoots, usually referred to as semi-mature, or half-ripe, because they are past the first flush of growth, but have not yet developed the woodiness of maturity, can be removed from many alpines and used as cuttings. These cuttings may be severed at a node as stem cuttings, or be taken as basal or heel cuttings (p. 69), depending on the material available—and more often than not, one kind of cutting produces roots as easily as another. They are usually snippets, 2.5–5 cm (1–2 in) long, with the leaves removed from the lower third of their stems, before being set up in a cutting compost (p. 76) in small pots. The removal of leaves in this way is customary, and is known as 'making' a cutting. Cuttings of plants with numerous small, or needle-like, leaves, such as alpine phloxes and saxifrages, are tedious and extremely troublesome to make, and it is hard to avoid inflicting severe injuries on them as it is done. Idle gardeners who grow bored and irritated by this fiddly business stick the cuttings in—leaves and all—and get results which are at least as good as those of their more dedicated brethren who conscientiously remove every leaf.

These cuttings should be taken as early as possible during summer to give them time to produce roots and develop into small established plants before the winter comes. Usually they are obtained from the replacement shoots which develop after plants have flowered and their emergence can be encouraged by lightly clipping over old plants as their flowers fade to remove seed heads, whose development would waste the resources of the plant. Occasionally, and in the case of kabshia saxifrages usually, it is worthwhile placing a pot over the crown of the plant, to shade it and draw up the shoots a little as they grow. These cuttings need to be hustled a little, and it is worth keeping them warm and well-tended to nurse them along while they are forming roots and becoming established. If they move fast they will make small plants by autumn with enough young shoots to develop well and flower prolifically the following summer. If the cuttings are taken a few weeks later than they need be or they suffer setbacks, they develop much more slowly and will probably not be ready to pot up until the following spring; their first flowers appearing a full year later than those of the early birds.

'Made' and 'unmade' cuttings of armeria, phlox, dianthus and saxifrage

Sixteen cuttings of an
encrusted saxifrage set up in
a 9 cm square plastic pot

Twenty cuttings of a dianthus
(pink) set up in a 9 cm square
plastic pot

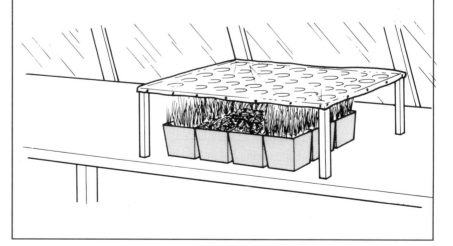

Alpines Propagated from Cuttings of Semi-mature Shoots

Acaena
Achillea argentea
  A. chrysocoma
  A. 'King Edward'
Aethionema 'Warley Rose'
Alyssum saxatile
'Compactum'
  A. s. 'Dudley Neville'
  A. s. 'Plenum'
Androsace lanuginosa
Arabis 'Rosabella'
Artemisia piedmontana
  A. schmidtiana 'Nana'
Aster alpinus
Aubrieta 'Bressingham
Pink'
  A. 'Dr Mules'
  A. 'Mrs Rodewald'
Campanula carpatica
'Isobel'
  C. c. 'Turbinata'
  C. garganica
  'W. H. Paine'
  C. pilosa
  C. zoysii
Cerastium columnae
Chrysanthemum
hosmariense
Cyananthus
Dianthus alpinus
  D. 'La Bourbrille'
  D. 'Mrs Holt'
  D. 'Oakington Beauty'
Diascia 'Ruby Field'
Dryas
Edraianthus
Erigeron glaucum
  E. 'Six Hills'

Erysimum 'Constant
Cheer'
  E. 'Jacob's Coat'
  E. 'Moonlight'
  E. 'Wenlock Beauty
Euryops
Gentiana × hascombensis
  G. septemfida
Gypsophila repens
'Dorothy Teacher'
Helianthemum 'Amy
Baring'
  H. 'Ben Moore'
  H. 'Lucy'
  H. 'Raspberry Ripple'
  H. 'Voltaire'
  H. 'Wisley Pink'
Hypericum olympicum
'Citrinum'
  H. o. 'Grandiflorum'
Iberis 'Little Gem'
  I. sempervirens
Lithospermum 'Grace
Ward'
  L. 'Heavenly Blue'
Micromeria
Moltkia
Penstemon pinifolius
  P. roezlii
  P. scouleri
Phlox amoena
  P. douglasii
  P. stolonifera
  P. subulata
Potentilla aurea 'Flore
Plena'
  P. tonguei
  P. verna 'Nana'

Saxifraga
Scabiosa
Sedum cauticolum
  S. heterodontum
  S. roseum
  S. spurium
  S. 'Vera Jameson'
Shortia
Silene maritima 'Flore
Plena'
Tanacetum densum ssp.
amani
  T. petralitum
Teucrium subspinosum
Thymus citriodorus 'Silver
Queen'
  T. herba-barona
  T. nitidus
  T. 'Peter Davis'
  T. 'Porlock'
Verbascum 'Laetitia'
Veronica 'Crater Lake
Blue'
  V. 'Mrs Holt'
  V. prostrata 'Loddon
Blue'
  V. 'Red Fox'
  V. 'Shirley Blue'
  V. spicata
  V. teucrium
Viola 'Ardross Gem'
  V. cornuta
  V. 'Hunterscombe
Purple'.

Figure 7.2   Cuttings prepared from semi-mature shoots provide a simple way to propagate many different kinds of alpines during the summer. Here the cuttings are of Armeria, Phlox, Dianthus and Saxifraga. In each, the right-hand cutting has been 'made' in preparation for setting it into cutting compost, by removing leaves and buds towards the base of the cutting. Cuttings of most kinds of alpine plants will produce roots most readily when provided with light, airy conditions, either in a well-ventilated cold-frame or set out on a bench in a greenhouse with a light overhead covering.

Propagation of Alpines from Rosettes and Offsets

Compressed shoots on which the leaves grow in close-set whorls are commonly found amongst alpines. Others produce clearly defined stems, sometimes only 1–2 cm ($\frac{1}{2}$–$\frac{3}{4}$ in) long, but sometimes much longer, on the end of which clustered leaves around a short stem form an offset or runner. Whether rosette or offset, shoots of this sort should be regarded by gardeners as cuttings waiting for roots to happen.

Stems with their clustered leaves should be cut as close to the parent plant as possible, and the lowest, oldest and dead leaves trimmed off before implanting the base in cutting compost. Usually they will do no more than rest on the compost, just lodged in place by the vestige of stem at their base. The closely compact form of these plants and their natural capacity to withstand exposed conditions makes it unnecessary to enclose them in frames or propagators while they produce roots, and to do so risks losses from decay. But all plants with this compact growth form do better when they are protected during their first winter, either on a bench in an unheated but lightly shaded greenhouse, or in a cold frame, in which the lights are permanently wedged open to provide ventilation; not to cosset them from the cold, but to protect them from lethal wet which creeps between their closely set leaves. The cuttings should be fed when roots appear, and potted up as soon as possible afterwards. Some of these plants, house leeks (*Sempervivum*) in particular, have a reputation for a frugal, self-denying life style, and are customarily expected to flourish with very little encouragement. However, even these respond to kindness, and should be fed like any other cutting when they produce roots.

Alpines that can be Propagated from Rosettes and Offsets

Androsace chumbyii	*Armeria caespitosa*	*Raoulia*
A. sempervivoides	*A. corsica*	*Saxifraga*
Arabis caucasica variegata	*A. maritima*	*Sempervivum*
A. ferdinandii-coburgii variegata	*Douglasia*	*Soldanella.*
	Draba	

Alpines Propagated from Wedge Cuttings

A few plants which are sometimes grown as alpines produce shoots clustered around the top of a substantial rootstock. Leaves emerge from buds with no discernible stems, and each winter the shoots, such as they are, retreat into the security of these buds, appearing as mere protuberances around the crown of the root. At first sight it would appear that nothing is there from which to form a cutting. Nevertheless, cuttings can be fashioned quite successfully by scraping away any layer of soil which might veil the crown, and then removing the shoots with a sharp knife, each with a small wedge-shaped base to support it cut from the rootstock below. These compact, but complete, equivalents to a heel cutting provide a means of increase when stocks are very low; most of the plants which produce them can be propagated more easily by division if large plants are available.

These wedge cuttings do not share the innate facility of rosettes or offsets to produce roots, and require much more care during the more prolonged period that they need to establish themselves. Cuttings taken in early autumn can be set up on the open bench of an unheated greenhouse, or in a cold frame, and will form roots during the winter and develop into small plants the following spring. Great care must be taken not to insert them too deeply in the compost, nor to overwater them, and always to provide them with every bit of ventilation that is possible short of letting them become so cold that they freeze solid.

Alpines that can be Propagated by Wedge Cuttings

Erodium	*G. sanguineum*
Geranium cinereum	'Lancastriense'
'Apple Blossom'	*G. subcaulescens.*
G. c. 'Ballerina'	

Propagation from Leaf Cuttings

Tropical plants provide many opportunities to use leaves as the source of regenerative tissues from which to produce young plants. This is a most unusual situation in plants from temperate parts of the world, and, as if to confirm their exceptional situation, the two temperate plants in which leaves form the main part of a cutting belong to a family whose other members are almost exclusively found in tropical regions. Even in these, as will be seen later, the leaves themselves are not the part which produces the new plant, and the term 'leaf cutting' is really a misnomer.

Ramondas and haberleas are these odd plants out. These two closely related members of the family Gesneriaceae resemble African violets with their clusters of flowers from the centres of rosettes of overlapping, persistent leaves joined to the crowns by quite long petioles. They branch to form new crowns infrequently, and possess no other structures which resemble anything usable as a cutting. Like their cousins the African violets, both can be propagated from cuttings made from their leaves and petioles.

It is a rule amongst broad-leaved plants that the position where a leaf is joined to a stem always harbours the site of a bud. These so-called axillary buds are conspicuous in many plants, vestigial in others, but in all cases their outgrowth to form shoots or flowers is a normal feature of development. Ramondas and haberleas both produce almost invisible axillary buds and only a tiny proportion develop into sideshoots. Those that do not, however, still retain their capacity to grow into new shoots under favourable conditions. Their opportunity comes when individual leaves are detached complete with the petioles at their base, and inserted into a cutting compost.

The leaves that make up their rosettes are closely packed together, and great care is needed to make certain that each petiole is detached at the point where it joins the rootstock. If it is broken off above that point the regenerative cells at its base will be left behind, and with them the tissue from which a new plant would be formed. The leaves and petioles are embedded in cutting compost by inserting the petioles at an acute angle, so that their bases are just below the surface, and they are then put into a well-ventilated frame or propagator in sheltered but not closely enclosed conditions. These leaves are quite remarkably capable of enduring desiccation and are far more likely to decay than die of drought. The plants grow naturally from crevices in cliff faces, and during the summer often shrivel up completely from lack of water. They are able to recover their original form and colour from the most sadly withered remnants, and leaf cuttings display a similar resilience.

Normally roots develop first, and may be produced even by broken petioles which lack their axillary buds, providing a false hope of success. Later, leaves are formed from the minute traces of the axillary buds, and from these, aided by regular feeds, rosettes slowly develop over several months to produce new plants. The young plants should be potted up using the smallest possible container into which they can be fitted, but not until they have produced well-established, fully formed rosettes and root systems.

Propagating Alpines from Root Cuttings

There are a number of alpines which can be propagated by making cuttings from their roots (p. 111). Some of these are difficult or slow to reproduce by any other method, but most can be propagated easily from their seeds. However, seeds are not a satisfactory way to propagate specially selected forms, or cultivars, and root cuttings provide a faster and more satisfactory way of increasing these than other methods of vegetative reproduction.

Genera in which Species or Cultivars can be Propagated from Root Cuttings

Anemone	*Erodium*	*Papaver*
Carlina	*Morisia*	*Pulsatilla*.

Division

Multiplying Alpines by Division

Many alpines, sprawlers or mat-formers that cling close to the ground produce roots naturally from their stems as they grow. Others form clustered colonies, made up of numbers of rootstocks, each with its subsidiary roots, crowned by a cluster of buds or shoots. Propagation consists of dividing the plants into sections or individual crowns, each already complete with stems and roots, and potting them up. Nothing could be simpler. Division is a very easy way to produce plants, but it is almost always worth the trouble of giving newly made divisions a little extra care under cover in a sheltered place for a week or two while their roots develop to establish strong young plants. Divisions that are planted straight into the garden or set out in nursery beds take time to become established and are very vulnerable to even a short period of drought, or cold or drying winds, let alone to more direct action from scratching cats or tweaking birds.

Alpines Propagated by Division

Acaena buchananii
  A. microphylla
Achillea atrata
  A. tomentosa
Androsace sempervivoides
Antennaria aprica
  A. dioica
Arenaria balearica
  A. caespitosa 'Aurea'
Artemisia piedmontana
Aster alpinus
  A. natalensis
  A. subcaeruleus
Campanula 'Birch Hybrid'
  C. cochlearifolia 'Miss Wilmot'
  C. 'Miranda'
  C. portenschlagiana
  C. pulla
Carlina
Cerastium tomentosum
Dryas octopetala
  D. suendermannii
Gentiana acaulis
  G. sino-ornata

Geranium cantabrigense
  G. cinereum
  G. orientalitibeticum
  G. pylzowianum
  G. sanguineum
  G. subcaulescens
Lewisia cotyledon
  L. rediviva
  L. tweedyi
Mazus
Ourisia
Parochetus
Paronychia
Phyteuma
Potentilla alba
  P. cuneata
  P. nitida
Polygonum
Raoulia
Saxifraga 'London Pride'
  S. 'Primulaize'
Scabiosa
Scutellaria
Sedum album
  S. cauticolum
  S. kamtschaticum

S. oreganum
S. sieboldii
S. spathulifolium
S. spurium
Sempervivella
Sempervivum
Shortia
Sisyrinchium angustifolium
  S. bermudianum
  S. macounianum
Soldanella
Solidago brachystachys
Tanacetum
Thymus 'Doone Valley'
  T. drucei
  T. lanuginosus
  T. necefferi
Veronica prostrata 'Alba'
  V. 'Mrs Holt'
  V. 'Rosea'
Viola biflora
  V. cucullata
  V. jooi
  V. labradorica
  V. odorata.

Propagation from Mat-forming Alpines

The most easily propagated alpines are the mat-formers which root as they grow, like thymes, veronicas and some of the polygonums; or plants which produce the root-like stems which are known as rhizomes.

Most mat-formers can be cut up to make a number of separate plants at almost any time of the year. However, their growth is very seasonal, and far better results will follow if this is taken into account. They grow and produce roots most actively during the summer and will rapidly form well-established young plants if pulled to pieces in June or July. Their natural tendency to produce roots can be encouraged by top-dressing them during the early summer with a shallow, partial covering of grit and peat; by the time they are divided about a month later they will already have started to produce masses of young shoots and roots, ready to be turned into plants.

Propagation from Rhizomatous Alpines

Rhizomes may be defined as stems which behave like, and often superficially resemble, roots. Some grow entirely underground, others snake along on the surface of the soil; and they take on a variety of disguises, several of which play a very significant part in plant propagation and will be referred to in later chapters. Horribly familiar examples, known to all gardeners, are the underground shoots of ground elder and couch grass. However, rhizomes are worth a closer acquaintance; when present they always provide a ready way to propagate a plant extremely easily. At first glance they very often look like roots, but the give-aways are the vestigial remains of the leaves and nodes, possessed by all stems, which reveal their true nature. These remnants are usually quite readily visible as scales or small brown bracts repeated at intervals on the surface of what may look like a root. But, roots are smooth and uniform throughout their length and never have these surface features. As is the rule, axillary buds occur at the positions where the leaves, vestigial though they may be, meet the stem. These buds are potentially active, and able to grow to form stems; so much so that plants which produce rhizomes are invariably easy to propagate and can be aggressively invasive, including such terrors as the giant knotweed, the willow herb, and the two examples referred to earlier.

The simplest way to propagate a rhizome is to cut it into sections each of which carries a few buds along its length. These, as all who have battled with couch grass learn quickly and bitterly, will produce roots rapidly and grow into plants. The best times to divide garden plants of this kind are either as they renew growth in the spring, or in late summer when they are starting to make preparations to survive the winter. The divisions should be potted up individually into very small pots, or set out, at the same sort of spacing as pricked out annuals (p. 59), in seed trays. They can be grown on in a cold frame to protect them while they become established, and then planted out in the garden as soon as they are able to fend for themselves.

CHAPTER EIGHT

Propagation of Hardy Perennials, including Grasses and Ferns

The term herbaceous perennial covers a broad assortment of plants which grow actively during spring and summer, producing leaves, flowers and seeds on annually produced stems. These die back to ground level in the autumn, and the plants survive during winter as an assemblage of resting buds at, or just below, ground level. They very often produce densely tangled mats of roots and crowns, which are persistent, and enable the plants to hold their ground and survive in competition with the vegetation around them. An important minority are evergreen, or nearly so, and remain not only visible, but often decoratively visible and active throughout the year. Many of these evergreens are natives of areas of deciduous woodland, where they survive by avoiding competition with the dense overhead canopy of leaves during the summer, and by their ability to make a living during the winter and early spring. Others are silver-leaved plants that make little growth in their native lands during the hot, dry months of summer, but remain active, and grow whenever conditions are favourable at other times of the year.

A high proportion of herbaceous perennials produce seed abundantly from year to year, and seed provides an extremely effective way of raising large numbers of plants rapidly at very little expense. Many of those grown in gardens are natural species which reproduce true to type from seed; others are selected strains or forms whose seedlings include high proportions of attractive plants. Some are extremely easy to germinate; however, because they are perennials and vegetative methods of propagation usually provide simple alternatives, germination responses have not been decisive factors in the selection of those which became cultivated plants, as is the case with annuals, and many herbaceous perennials do possess quite complex germination requirements. Herbaceous perennials occur in a variety of situations but particularly those in which woodland, or scrub merging into grassland, form the natural vegetation. These natural associations with woodland and grasses still play a part in the ways they behave as garden plants and should be taken into account when we propagate them.

These perennial plants survive by occupying a space and competing successfully with usurpers. They must be capable of maintaining their hold through the winter, and, even more critically, be able to grow away strongly the following spring, even if the weather is unfavourable. Most manage this by accumulating storage reserves, particularly of starch, in their crowns, main roots or rhizomes during late summer and autumn to maintain respiration through the winter, and support renewed growth the following spring. Special storage organs such as tubers are occasionally formed and, because these are conspicuous, make it all the easier to overlook the fact that plants which do not possess them also have very substantial reserves stored in their normal unspecialised tissues. Successful vegetative propagation takes account of these reserves, not only in ways which avoid wasting them, but by giving newly propagated plants adequate opportunities to lay them down. Parts of the plants removed as cuttings or divisions during late spring, immediately before the renewal of growth, are likely to be well-stocked with starch and other reserves, and able to make immediate effective use of them while they establish themselves. The traditional practice of digging up and dividing plants in the middle of the winter pays no attention to cycles of growth and development in these plants. It may be convenient, and many perennials are so tolerant that they can survive disturbance and severe damage, even at this time when their growth processes are extremely subdued. Less tolerant varieties and species have difficulty in repairing damage to roots and crowns; they may establish with some difficulty, or even die before they do so. Periods of severe weather at this season increase the risks of losses enormously.

As with leaves, flowers and stems, the roots of herbaceous perennials, hidden beneath the ground, also go through annual cycles of growth and decline and renewal. Many perennials share a pattern of root development which starts in late summer with the growth of strong, downward-tending anchor roots, produced at the base of newly formed buds or crowns, and these penetrate the soil deeply to provide a secure foundation for the winter. These roots then become filled with starch. Their importance as storage organs is equally as vital to the plant as their anchoring effect. The following spring the storage reserves in the roots are broken down and the energy stored in them from the previous year is used to support the actively developing young shoots; at the same time a network of feeding roots is produced from the anchor roots and from the base of the crowns. These explore the surface layers of the soil around the plants, providing the developing shoots and flowers with nutrients and water. As the summer progresses, and the need to sustain rapid growth becomes less vital, the major part of the root system wastes away and disintegrates to be replaced when the cycle starts again in late summer. Many of these plants can be propagated very effectively by division during July and August, at the start of this annual cycle, even though it is a time when they may be heavily encumbered with foliage and will frequently be in flower.

Seed Germination

Easily Grown Perennials

Many herbaceous perennials produce seeds which are not at all difficult to germinate, and can be sown and treated like the annuals described previously (p. 56). All are hardy and almost all can be sown and grown without any artificial warmth. However, they produce seedlings faster and make more rapid growth when they are warm rather than cold, and this makes it worthwhile maintaining temperatures of 15–20°C (59–68°F) if possible. A high proportion of the species involved grow naturally in parts of the world where grasslands prevail, and are adapted to germinate rapidly, and develop into small plants quickly under very competitive conditions (p. 18).

Perennials that Respond to and Benefit from Warm Conditions

(All can be sown at 15–20°C (59–68°F), under cover, between January and April.)

Acanthus	Festuca	Pennisetum
Adenophora	Filipendula	Penstemon
Althaea	Foeniculum	Phormium
Anaphalis	Fragaria	Phygelius
Anchusa	Gaillardia	Phytolacca
Aquilegia	Galega	Platycodon
Baptisia	Gentiana	Polemonium
Briza	Geranium	Potentilla
Catananche	Geum	Primula
Centaurea	Glaucium	Pterocephalus
Cephalaria	Gunnera	Pyrethrum
Cerastium	Gypsophila	Ranunculus
Chelone	Heuchera	Rhazya
Chrysanthemum	Humulus	Rheum
Codonopis	Hyssopus	Rodgersia
Cortaderia	Incarvillea	Romneya
Cynara	Lupinus	Rudbeckia
Delphinium	Lychnis	Ruta
Dianthus	Macleaya	Salvia
Diascia	Malva	Sárracenia
Dicentra	Meconopsis	Scabiosa
Digitalis	Melianthus	Sidalcea
Dodecatheon	Mimulus	Stachys
Doronicum	Monarda	Tellima
Echinops	Morina	Tiarella
Erigeron	Nepeta	Verbascum
Eryngium	Oenothera	Viola.
Eupatorium	Panicum	
Euphorbia	Papaver	

An alternative method of raising seedlings of many herbaceous perennials, especially those that come from areas of the world where they are associated naturally with grasses, is to sow them during August in containers. They will germinate before the autumn and pass the winter, protected by a cold frame or unheated greenhouse, in their original containers. The young seedlings can be pricked out or potted up individually when they start to grow early the following spring. Their roots may be rather entangled when the time comes to prick them out, but can be separated easily and without damage by immersing the whole root ball in water, and gently easing the seedlings apart.

Hardy Perennials that can be Sown in July and August and Overwintered as Seedlings

Adenophora	Festuca	Omphalodes
Althaea	Fragaria	Physalis
Baptisia	Geum	Phytolacca
Bellis	Glaucium	Platycodon
Campanula	Gypsophila	Polemonium
Catananche	Incarvillea	Potentilla
Centaurea	Lathyrus	Primula
Cephalaria	Liatris	Pyrethrum
Chrysanthemum	Limonium	Rheum
Codonopsis	Linum	Salvia
Delphinium	Lupinus	Saponaria
Dianthus	Lychnis	Scabiosa
Digitalis	Malva	Tiarella
Dodecatheon	Mertensia	Verbascum
Doronicum	Morina	Viola.

Sowing Perennials Out of Doors

A great many herbaceous perennials are strong, enduring plants, which produce seedlings under very simple conditions, and not only survive, but flourish, with no mollycoddling and very little attention, apart from being kept free of weeds. Plants of this kind can be sown outdoors, during spring and early summer in drills in a nursery bed. An early start is not necessary and it is better to wait until May or early June to allow the soil to warm up before sowing any seed. The young plants can be left in their nursery beds until the late autumn and planted out in their permanent positions some time between September and March.

Hardy Perennials that can be Sown Outdoors in Spring or Summer

Anaphalis	Crepis	Foeniculum
Anchusa	Delphinium	Gaillardia
Aquilegia	Digitalis	Galega
Armeria	Echinops	Gypsophila
Centaurea	Erigeron	Hesperis
Centranthus	Eryngium	Heuchera
Chrysanthemum	Eupatorium	Hyssopus
Crambe	Euphorbia	Iberis

Lamium	*Onosma*	*Salvia*
Liatris	*Papaver*	*Saponaria*
Limonium	*Penstemon*	*Scabiosa*
Linaria	*Phygelius*	*Senecio*
Lunaria	*Physalis*	*Sidalcea*
Malva	*Plantago*	*Thalictrum*
Meconopsis	*Potentilla*	*Tiarella*
Milium	*Pterocephalus*	*Verbascum*
Oenothera	*Ruta*	*Viola*

Perennials that Require Conditioning Treatments

The hardy plants which grow naturally in deciduous woodlands or in scrub amongst shrubs very frequently have more complicated germination responses. Their seeds are more likely to possess means of postponing or controlling the times when seedlings emerge, and these must be reckoned with when they are sown.

A great many seeds possess a very straightforward way of postponing germination: freshly shed seeds are incapable of germinating, but, more often than not, this inability is lost spontaneously, after the seeds have been kept dry for a few weeks in a process known as after-ripening. Seeds of this kind should not be sown immediately after being harvested, nor, if they are going to be stored should they be hurried into a refrigerator or freezer, because the changes which enable seeds to germinate are delayed, or prevented, by low temperatures.

The seed coats of some seeds are extremely hard and thick, and may also be heavily impregnated with waxes and vegetable fats making them completely impermeable to water. The embryos inside these seeds remain consistently dry until the seed coats eventually disintegrate sufficiently for water to seep through to them. Under natural conditions these barriers are gradually broken down by fungi or bacteria, or damaged mechanically by a rodent's teeth or a bird's beak during unsuccessful attempts to eat the seeds. These produce cracks through which water enters the seeds, and once this happens germination usually occurs without further delay. These seeds are mostly extremely easy to germinate once the problem is diagnosed and their seed coats deliberately damaged (p. 50).

Another device which ensures that germination does not take place immediately is the production of seeds in which the embryos are only partially formed when the seeds ripen. The buttercup family is particularly expert at this; their immature embryos only become capable of germination when the seed is hydrated, and so they do not develop when seeds are being stored under dry conditions. The embryos develop most satisfactorily at 10–20°C (50–68°F), corresponding, very broadly, with the range of temperatures that might be expected immediately below the soil's surface in temperate regions during late summer and early autumn. Many of the plants which produce seeds of this kind are woodland species and grow in places where the soil is likely to become rather dry during the summer because of the combined effects of the umbrella-like leaf canopy overhead, and the presence of tree roots below. The possession of immature embryos not only prevents premature germination in this uncertain situation, but provides a way of responding to moist conditions whenever they do occur. Depending on their situation, and the conditions which control the rate at which the embryos develop, seeds of this kind will probably produce seedlings sometime during the autumn or winter after they are shed.

Conditioning treatments at low temperatures similar to those used for seeds of alpines (p. 88) also improve the germination of many herbaceous perennials. The effects of these chilling temperatures are quite familiar: much less well known are conditioning treatments of a directly opposite kind. These are found amongst species which germinate best in cool situations after their seeds have been conditioned at high temperatures, from 20–30°C (68–86°F). The range and variety of such species is very wide, but many species which may react in this way remain unknown. It is not easy to distinguish species which need to be conditioned at high temperatures from those with immature embryos. In both cases, however, seeds will lie without germinating when temperatures are high, but, provided they are kept moist, they will start to germinate after a period of several weeks, or months, usually at lower temperatures, equivalent to those of late autumn or winter.

As soon as they ripen, seeds of all such plants can be sown in containers set out on a bench in a greenhouse. In the autumn they should be transferred to an unheated frame in which the lights are closed only during periods of frosty weather, to give them cover, but to expose them to the normal change of temperatures as the seasons progress. Provided they experience sufficiently high temperatures during the first few weeks after being sown, their seedlings should start to emerge during the late autumn or early the following spring.

Although it is simple and convenient to follow the natural progression of the seasons, this is possible only when seed is available to be sown during the summer. An alternative is to mix seed with moist vermiculite in a transparent polythene freezer bag, as described earlier (p.89), and to put it somewhere warm to condition—the hot water tank in the airing cupboard, a heated propagating frame, or an electric propagator may all be used. The high temperatures should be kept going for at least six to eight weeks before the seeds are moved to cooler conditions, the best place being in a refrigerator for four to six weeks. They can then be spread out over the surface of compost in containers and set out on the bench of a cool greenhouse to germinate. However it is achieved, these seeds require time to produce their seedlings, and this needs to be taken into account so that when seedlings do emerge they have enough time ahead of them to establish well-based young plants before they are subjected to the hazards of a winter.

Genera that are Likely to Respond to High Temperature Conditioning Treatment

Aconitum	Astrantia	Epimedium
Actaea	Bupleurum	Eranthis
Anemone	Caltha	Erodium
Anthriscus	Cimicifuga	Eryngium
Aquilegia	Dictamnus	Gentiana

Geum	*Ligularia*	*Pulsatilla*
Helleborus	*Myrrhis*	*Trollius*
Heracleum	*Peonia*	*Veratrum.*

*Included in this list are some plants which respond because their seeds contain immature embryos.

Propagation of Ferns from Spores

Life Cycle of a Fern

Ferns never produce flowers; however, during the late summer, a glance at the undersides of the fern's fronds will reveal masses of scale-like structures, like nothing to be seen on leaves, covering a high proportion of their surfaces. As these develop they become brown, and split open to release quantities of snufflike spores. It would be easy to assume that these are the fern's equivalent to seeds.

They are—in a way—but really they are not. Fern spores resemble seeds in that they provide a way for plants to move around and seek out new locations in which to grow; they are also part of the process of sexual reproduction. Contrariwise, they are unlike seeds in that they represent quite a different part of the plant's life cycle. Seeds are a way of packaging plant embryos after fertilisation, during which the genes of the parents combine to form a newly constituted plant. Fern spores are simply genetic replicas of the plant which produces them; the process of genetic recombination is still in the future. They do provide a very effective way of producing fern plants but only if the ways that they grow and develop are understood and catered for.

Biologically, ferns are very different from, and exceedingly remote relatives of, flowering plants. The conspicuous fronds, crowns and rhizomes which make up a fern plant may look like the leaves, stems and roots of more familiar plants, but, in fact, represent quite a different stage in their life cycle.

Flowering plants reproduce sexually by seeds, the result of fertilisation of pollen and ovules. Each embryo contains a half share, and a random half share, of the genes of its two parents, and each is usually different from its siblings which inherit various other combinations of their parents' genes.

Fern spores are the last fling of the old generation. They are a form of vegetative reproduction by which cells in the fronds divide and multiply to produce thousands of spores, each one genetically identical to its parent and to all its brethren. These spores drop off and drift away in the air currents and, if they settle in a suitable spot, resume their growth and develop into small, flat, green objects, rather similar to liverworts. These insignificant structures, each known as a prothallus, are the fern's equivalent of the plants which flower in our gardens.

As they mature, and this does not usually take long, the prothallii produce separate male and female organs on their surfaces, and within these are either small mobile spores (the equivalent of pollen) or relatively large, sedentary cells

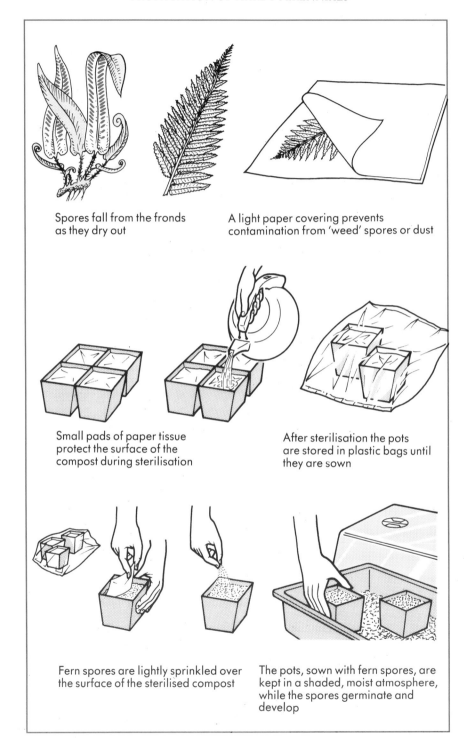

Spores fall from the fronds as they dry out

A light paper covering prevents contamination from 'weed' spores or dust

Small pads of paper tissue protect the surface of the compost during sterilisation

After sterilisation the pots are stored in plastic bags until they are sown

Fern spores are lightly sprinkled over the surface of the sterilised compost

The pots, sown with fern spores, are kept in a shaded, moist atmosphere, while the spores germinate and develop (representing the ovules). The numerous mobile cells can only move about and come into contact with the sedentary cells if there is a film of water covering the surface of the prothallus. Once contact is made the nuclei of the two cells fuse to produce an embryo. The fertilised ovules are not enclosed in anything that resembles a seed, but remain embedded in the tissues of the prothallii, which support and nourish them during the early stages of embryo development. Eventually, if they survive, these grow up to produce the familiar ferns, and the prothallii, having served their purpose, shrivel up and disappear.

   Young fern plants, developing from the embryo and nourished by the prothallus, are the equivalent of seedlings, and mark the start of a new generation. They are the product of a sexual process, and differ genetically from their parents and from each other. However, unless spores from different plants produce prothallii which grow extremely close to each other, the opportunities for cross fertilisation from one plant to another are very limited, and the offspring are more likely to be genetically similar to their parents than are the seedlings of flowering plants. Ferns give the impression of being unlike other plants—and they are. Many, in their mature form, are particularly well adapted to survive periodically inhospitable conditions, either from drought or from dense shade. All of them, during their time as prothallii, are dependent on continuously high humidity, and the presence of a film of water to function properly. Mature ferns become securely entrenched and are typically very long-lived, and well able to resist competition from other plants; the prothallii exist for only a few weeks, and are extremely vulnerable to competition, not only from higher plants, but especially from the mosses and liverworts which share their enjoyment of constant moisture, and coolly shaded corners.

Collecting Fern Spores

Most ferns produce their spores on the undersides of their leaflike fronds in ranks of conspicuous spore cases. A few, like the royal fern, produce them on fronds specially modified for the purpose, which bear no resemblance to a leaf but develop an upright framework forming a tracery of filaments beset with clusters of spore cases. As the spores mature they become brown and desiccated and the cases which hold them split to release them into the air. Spores can be collected at this stage by removing the fronds, and laying them down over sheets of plain white paper in an unused room, or some other dry, well-ventilated but draught-free place. A single sheet of face tissue placed over the fronds will protect them from dust and prevent contamination with other spores, particularly those of their arch rivals the mosses and liverworts. As the fronds dry out, they release their spores and these fall onto the paper below, forming in the process a shadowy representation of the frond from which they came.

   After a few days the spores can be collected and sown. This is not difficult but different, and will be successful only if the special features of a fern's life cycle are understood and provided for. Collection is quite straightforward, but for

---

Figure 8.1   Ferns can be grown from the spores which are produced on the under-surfaces of their fronds.

one thing. The spores are minute, barely visible and dust-like. The spore cases which enclose them are relatively large, and easily become detached and mixed with the spores. The fronds should be handled very gently to try to avoid this happening, and later, when sowing the spores, great care must be taken to make sure that it is the spores which are being sown, and not their useless cases. A magnifying glass is a great help.

Sowing and Growing On Fern Spores

Ferns can be grown from their spores, often without great difficulty—though these ancient plants are not to be hurried, and the process takes some time. Many produce spores which can be stored successfully in the same way as small, dry seeds of flowering plants (p. 181). However, the spores of some species are not long-lived and it is always advisable to sow a part, at least, of those collected as soon as possible. The most vital step is to make sure that they have the conditions they need during the critical stages and vulnerable period when they exist as prothallii. There are two things that must be attended to, otherwise failure will be the only result.

- The first is to keep them in surroundings which are constantly moist and cool, from the time that the spores are sown and the prothallus develops, until after the tiny fern plants appear and become independently established.
- The second is to prevent the competitive growth of mosses and liverworts, which are the weeds of this process, and can overwhelm the prothallii as well as the immature fern plants.

Ferns grow from spores; so do mosses and liverworts—and these unfriendly spores are in the air, and even more insidiously in water stored in tanks and garden ponds. The secret of success is to sow the fern spores on to a surface which has been scalded to destroy any alien spores already on it, and always to use clean water—preferably boiled tap water—to keep the prothallii moist as they develop.

A simple and effective way to do this is to fill a number of small square plastic pots with a peaty potting compost, and then sterilise the surface of the compost. This can be done by pouring boiling water from a kettle over pads of face tissue cut out to fit, and laid on the surface of the compost. The scalding water drains through the compost in the container, and when it has done so more should be added; to ensure a thorough effect this should be repeated at least twice more. As each pot is prepared it can be put temporarily into a polythene bag and set on one side.

The next step is to sow the spores. The bags containing the scalded pots are taken one by one and opened; and the face tissues on the surface of the compost are peeled off and thrown away. A pinch of spores is then scattered quickly over the surface of the compost and the polythene bag resealed without delay with a twist of paper-wrapped wire. The usual advice is to sow the spores evenly and moderately thickly, but they are so tiny that nothing will be visible, and it is impossible for mere mortals to gauge how densely they are distributed.

The fern spores are most likely to develop into prothallii in a cool place in diffuse light, and this is where the sealed bags should be put. It is particularly vital to make certain that it is not a place in the sun. Even a short period of direct sunlight can be lethal, and a completely shaded corner is needed, perhaps under a bench in a greenhouse, a corner of a well-lit shed, a little used room in the house, or a heavily shaded cold frame or cloche.

The spores will start to develop into prothallii in the humid conditions within the polythene bags. At first nothing will be visible, then, if all goes well, a green film spreads over the surface of the compost, and later the plate-like shapes of the prothallii become discernible. As these grow and develop it becomes possible to see the craters and pits on their lower surfaces in which the male and female cells are produced. During the whole of this time the atmosphere around the prothallii must be kept humid, but it is neither necessary nor beneficial for the compost on which they are growing to be soggy wet. Water should be applied only when it is absolutely necessary and must be boiled before being sprayed, as a fine mist, very lightly over the surface of the compost; a small hand-held scent spray is just right for the purpose.

When the prothallii mature, the male spores depend on a very fine film of water to migrate to the sedentary females, either on the same or neighbouring prothallii. With this in mind it is a temptation to overdo the watering with repeated sprays of boiled water; temptations, so we are taught, should always be resisted, and for once this is good advice. These prothallii may need humid conditions and a film of water, but they rot away very easily if they are kept constantly wet—the distinction is a fine one! Very sparing applications of water are all that will be needed, provided the polythene bags are well sealed. When all goes well the ovules, lying within their craters, will be duly fertilised and will start to grown into tiny fern plants. Minute fronds emerge, distributed over the surface of the prothallii, and often several will appear from each one so that before long the entire surface of the compost in each pot develops a haze of foliage. Once the small ferns can be seen the pots can be removed from the protection of the polythene bags and set out in a sheltered semi-shaded place in a greenhouse or frame where the plants can continue to develop. After several months, up to a year with slow-growing forms, the young clusters of develop-ing plants will be large enough to separate into individual plants or small entangled groups and set out in small pots, in which they can continue to grow until large enough to plant outside.

A very high proportion of the ferns grown in gardens have been selected from plants growing in the wild with unusual or attractive forms. Some have repeatedly sub-divided, very feathery fronds; in others the margins are wavy or the mid-ribs crested, or some feature is greatly reduced or exaggerated. These forms are very frequently genetically inherited, although they seldom come completely true when spores are sown. However, the mechanisms for gene exchange during sexual propagation are primitive and do not work very effectively in ferns, and quite high proportions of the offspring are likely to resemble their parents, and many of the remainder will develop into decorative forms, which are worth a place in the garden. The seeds of flowering plants may be a notoriously unreliable way to perpetuate selected varieties that take our fancy, but the spores of ferns offer rather more than a sporting chance of success.

Propagation from Cuttings

Propagation from Basal Cuttings

As hardy perennials start to grow in the spring a great many produce shoots which emerge from buds which have spent the winter just below ground level. These young shoots can be removed, and treated as cuttings, and very often produce roots rapidly and very easily. Indeed, if separated from the crowns just below the ground many will already possess the first traces, or even well-developed stages, of the feeding roots which they produce naturally at this time of year.

These basal shoots can be removed economically and with very little disturbance. Alternatively, the clusters of buds and roots, which make up the crowns or stools, can be dug up late in the winter, and either potted up, or set out in boxes and covered lightly with a shovelful of good garden soil, used compost or peat. If they are put into a cool greenhouse during February and March they will start to grow a little earlier than the plants left out of doors, and the cuttings they provide will have time, not only to produce roots, but to grow on into well-established plants that will flower during their first season. The parent plants used in this way should not be dug up too early. Most perennials become quiescent when they die down in the autumn, and will not grow satisfactorily until they have experienced a period of low temperatures, after which they burst into growth with renewed vigour as soon as temperatures rise a little. Those that are lifted too early, and that may be any time up to the middle of January, will refuse to grow satisfactorily even though large sums of money are spent on keeping them luxuriously warm.

Basal cuttings are cut off immediately below the point where they begin to feel firm to the touch. They can then be set up in small square plastic pots, in a cutting compost (p. 76), and kept in a humid atmosphere in a closed frame or propagator. Artificial heat maintaining minimum temperatures of 10–15°C (50–59°F) at the base of the cutting helps to reduce losses from damping off, and greatly increases the speed at which they produce roots and become established. Soon after they have produced roots they should be fed (p. 82) and then transferred to a cold frame. After a week or two they will be ready to pot up individually, when they will develop rapidly into small weather-tolerant plants that can be planted out in the garden early in the summer.

Perennials that can be Propagated from Basal Cuttings in the Spring

Ajuga	*Crambe*	*Geranium*
Anaphalis	*Delphinium*	*Gypsophila*
Anchusa	*Diascia*	*Helenium*
Aster	*Epilobium.*	*Humulus*
Campanula	*Erigeron*	*Lamiastrum*
Centaurea	*Erodium*	*Lamium*
Chrysanthemum	*Euphorbia*	*Linaria*
Codonopsis	*Gaillardia*	*Lupinus*

Lychnis	*Monarda*	*Scabiosa*
Lysimachia	*Nepeta*	*Scrophularia*
Lythrum	*Nymphaea*	*Sedum*
Macleaya	*Phlox*	*Solidago*
Malva	*Physostegia*	*Teucrium*
Mimulus	*Salvia*	*Verbena.*

Propagation from Rhizomes

Those hardy perennials which produce rhizomes can be propagated in a simple, satisfactory and effective way by cutting the rhizomes into sections and treating each section as a cutting. If this is done between the beginning of May and the end of June most establish themselves extremely rapidly and need very little skill and the minimum of care for a successful result. The root-like stems of rhizomes may grow underground, but many, like those of bergenias, snake along at ground level. They may be wide ranging, questing, attenuated and thin in cross section, or thick and stubby, advancing steadily, in stages measured only in centimeters each year. Whatever their shape or size, all rhizomes possess the hallmarks of all stems, and contain buds at set points along their length which provide the regenerative tissues needed to form a new plant.

Amongst the easiest of all plants to propagate are some aquatic and marginal water plants. Many of these produce strong, questing rhizomes in which the nodes are well marked, and roots frequently emerge as the plants grow, especially during spring and early summer. These can be cut into sections during May and June, potted up individually and kept, either just submerged in water or under almost saturated conditions according to their nature. They establish themselves seemingly in a matter of days, and are ready to take their place in the water garden before a month has passed.

Perennials that can be Propagated from Rhizomes

Aegopodium	*Epimedium*	*Nuphar*
Astrantia	*Geranium*	*Nymphaea*
Bergenia	*Helianthus*	*Physalis*
Campanula	*Macleaya*	*Polygonum*
Centaurea	*Menyanthes*	*Thalictrum.*

Propagation from Root Cuttings

The word 'cutting' is most likely to call to mind an image of a leafy shoot, most probably cut from a shrub of some kind. However, roots are the complements of shoots, and make up the cryptic, barely known and almost completely ignored aspect of the Janus head of plants, of which the flowers and shoots are the more familiar face. Roots contain cells which divide, and which can, if circumstances suit them, reproduce any part of a plant including shoots and flowers. Slice through the roots of a dandelion with a spade, and throw away the top: a month later, a perfectly healthy, multi-crowned dandelion is growing and flowering in its place, reconstructed in every detail from the roots that were

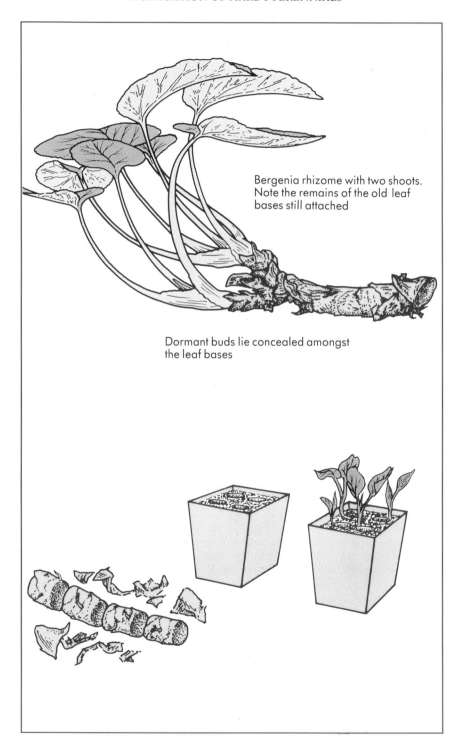

Bergenia rhizome with two shoots. Note the remains of the old leaf bases still attached

Dormant buds lie concealed amongst the leaf bases

left behind! The list of plants which may be propagated in this way is quite short; many more might be added once there is a greater understanding of the ways that roots grow and the times of year when cuttings are likely to succeed.

The anchor roots, produced by many perennials during late summer, are the key and provide the material, well-provisioned with stored starch, from which to prepare root cuttings during January and February, and into March, just before growth is due to start in the spring. It is a season when the roots are physiologically prepared to release their storage reserves, and their tissues are ready to grow actively as soon as they are provided with favourable conditions.

As soon as they are needed the parent plants are dug up, and some of their strongest roots are cut off close to the point where they emerge from the crown. No more than half the roots present should be removed, preferably only a third, and the donors should be carefully replanted as soon as they have been stripped of their roots. If necessary, it is possible to remove virtually every one of these large roots without irretrievably damaging the plants they are taken from. But when this is done, it is not sufficient simply to replant the parent and forget about it: the plant should be potted up and given a little extra care and cosseting in a sheltered place. It will soon recover, and can then be replanted.

The lengths of root are made into cuttings by dividing them into sections, each 2 or 3 cm ($\frac{3}{4}$–1$\frac{1}{4}$ in) long. It is essential to remember that each section has two ends, and one of those is the top, and one the bottom, depending on how they were positioned while on the plant. It is important not to disorientate them by inserting them upside down. The easiest way to avoid this is by laying them out carefully, as they are cut off, in orderly rows so that the upper ends lie pointing away from the operator.

The prepared cuttings are set out in cutting compost in containers; and pots which are deep enough to drain well are more satisfactory than shallow trays. The best procedure is to start by banking up one side of the container with a very gritty compost to make a slope at about 45° against which the first row of cuttings is placed. These are then covered with a layer of compost and another row set out, and so on until the container has been filled. The slices of roots can be placed close together, from 1-2 cm ($\frac{1}{2}$–$\frac{3}{4}$ in) apart depending on their diameter, and their tops should be immediately beneath the surface of the cutting compost. These cuttings will produce shoots more rapidly, and more successfully, if the containers are put into a frame or propagator fitted with thermostatically controlled heating cables which maintain temperatures of 10–15°C (50–59°F) in the compost around the cuttings. If no heated accommodation can be spared they should be placed in a sheltered protected place, such as a cold frame or unheated greenhouse.

Shoots are produced from the cut surfaces at the tops of the root cuttings, and

---

Figure 8.2 Some perennial plants can be propagated from their rhizomes. These root-like stems, sometimes entirely subterranean, sometimes spreading over the soil surface, contain nodes and buds, and are naturally inclined to produce roots abundantly. They can be used to produce new plants if cut into sections, each of which contains at least one bud. The two shoots can be cut off and used as cuttings. The remainder of the rhizome can then be sliced up into sections which should be set, right way up, into a gritty cutting compost. Dormant buds hidden amongst the leaf bases will then grow out to form shoots which develop into new plants.

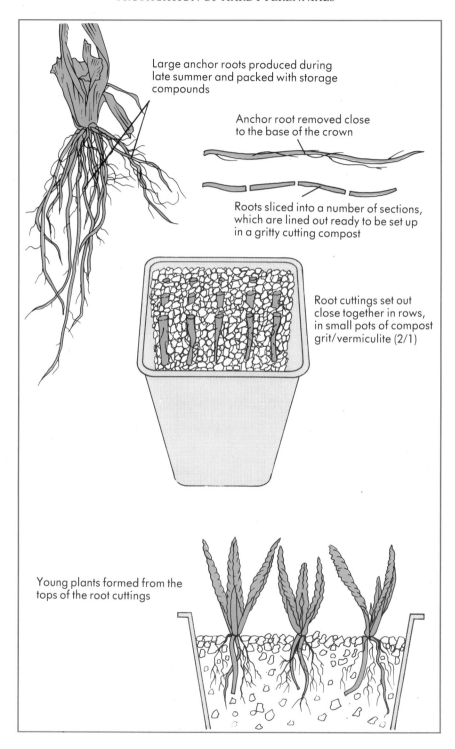

Large anchor roots produced during late summer and packed with storage compounds

Anchor root removed close to the base of the crown

Roots sliced into a number of sections, which are lined out ready to be set up in a gritty cutting compost

Root cuttings set out close together in rows, in small pots of compost grit/vermiculite (2/1)

Young plants formed from the tops of the root cuttings

should soon start to grow through the thin layer of compost covering them. Normally they will be followed by the appearance of feeding roots emerging, either from the original root cutting or from the area at the base of the newly formed shoots. Whichever way the cuttings develop they should not be disturbed until it is certain that the new roots have made enough growth to support the young developing shoots. Meanwhile they should be fed, like any other cuttings (p. 82), as soon as new growth shows that they are capable of making use of added nutrients.

Root cuttings provide a means of producing plenty of plants when other methods of propagation are very slow, or ineffective. However, the shoots they produce grow from very small beginnings, and are not likely to be large enough to plant into permanent positions in less than four to six months. They can be grown on through the summer in pots while they make this growth, and planted during the following autumn, provided time and inclination are available to give them the almost daily attention that this method of production demands. Alternatively they can be potted up into small pots soon after they produce roots, and, once they have established themselves as self-sustaining plants, can be planted out close together in rows in a sheltered nursery bed outdoors (p. 31). In most circumstances they will thrive better thus, and be ready to plant in the garden during the following winter or spring.

Border phloxes are frequently propagated from root cuttings, which may seem strange to those who know that basal cuttings of these plants root well. The reason is stem eelworm, an extremely serious pest which afflicts phloxes, and can totally wreck and quickly destroy susceptible varieties. Eelworms are harboured in the soil, and can remain alive and active in it for some years. They enter the plant at soil level and move up into the stem and multiply; but, being upwardly mobile creatures, they never descend into the roots. As a result, clean healthy new stock can always be obtained by growing these plants from root cuttings, provided that the young plants are isolated from any further source of infestation.

However, the story has yet another angle. A few varieties of border phlox have strikingly variegated leaves—'Norah Leigh' and 'Harlequin'are the best known. Eelworms will infest these like any other phlox, so their roots too could be cut up to produce young plants free from eelworm. The snag is that the young plants will produce leaves which are plain green and without a vestige of variegation, which rules out root cuttings as a way of reproducing these variegated kinds.

This underlines the fact that root cuttings almost always produce plants with normal green shoots; the technique is useless to propagate plants with variegated leaves. The reason is that variegation is commonly caused by a single cell in a growing point developing a very slight malfunction in one of the many steps involved in the production of chlorophyll. If this fault persists all the tissues

---

Figure 8.3   Perennial plants of many kinds produce large starch-filled roots, which can be cut up and used as a means of propagation during the late winter. The anchor roots of Primula denticulata *provide a means of reproducing selected plants with unusually large flowers, or particularly intense colours, which could not be relied on to come true to type from seed.*

which result from divisions of that cell inherit its inability to produce chlorophyll, and parts of leaves derived from it remain white or cream coloured. When root cuttings are used to propagate a plant, the cells within them possess the genetic constitution of the original green plant. Unless, by a remote coincidence, there is a repetition during cell division of the slight and very unusual error, all the new shoots will produce perfectly normal green leaves, in obedient compliance with the genetic information contained in their chromosomes.

Perennials that can be Propagated from Root Cuttings

Acanthus	Erodium	Pulsatilla
Anchusa	Eryngium	Romneya
Anemone	Gaillardia	Rudbeckia
Armoracia	Limonium	Scorzonera
Catananche	Macleaya	Tragopogon
Crambe	Papaver	Tropaeolum
Dicentra	Phlox	Verbascum.
Echinops	Primula	

Propagation from Semi-Mature Cuttings

According to definition the leaves and stems of herbaceous perennials die during the autumn or early winter, and the living parts of the plant hibernate just below the ground. But many hardy plants are not typical in this way and have persistent shoots which remain above ground all through the year. Many plants of this kind can be propagated from semi-mature cuttings taken during July and August, using methods which are similar to those described for some alpines (p. 91).

Perennials that can be Propagated from Semi-mature Cuttings During the Summer

Achillea	Dianthus	Iberis
Artemisia	Erodium	Linum
Calamintha	Geranium	Pterocephalus
Cerastium	Hyssopus	Veronica.

A large number of evergreen, permanent perennials can be propagated from cuttings taken a little later in the season, during September, and this is particularly useful as a way of ensuring the survival of species and cultivars which are not quite reliably hardy. These are treated in the way described for cuttings of silver-leaved shrubs (p. 140). They produce roots by early November, and can be overwintered safely in their containers and potted up the following spring. Once their roots have developed, the temperature of the heating cables can be turned down to no more than 5°C (41°F), which is just enough to protect the young cuttings from frosts.

Perennials that can be Propagated from Cuttings taken Early in the Autumn

Achillea	*Melianthus*	*Stachys*
Artemisia	*Onosma*	*Teucrium*
Diascia	*Origanum*	*Thymus*
Epilobium	*Penstemon*	*Verbascum*
Erysimum	*Phygelius*	*Verbena*
Euphorbia	*Polygonum*	*Viola.*
Geranium	*Ruta*	
Gypsophila	*Salvia*	

Propagation of Ferns from Leaf-base Cuttings

One of the toughest and most versatile of the hardy native British ferns is the hart's tongue (*Phyllitis scolopendrium*). A great number of varieties have been found at one time or another, some of which are still being grown as garden plants. The plants produce clusters of shining evergreen fronds from a very long-lived rootstock, and as the plants age, so the older fronds die away to be replaced by new ones. When the old ones wither away they do not quite disappear, but retain a memory of their existence in the form of the basal part of their fronds which persist around the crowns close to the ground. These frond-bases can be detached, by peeling them off one by one, and used to produce new plants by treating them in the way described earlier (p. 95) for leaf cuttings.

Division

Propagation of Hardy Perennials in the Winter

By long tradition gardeners have multiplied herbaceous perennials by digging them up sometime between November and March, dividing them into a number of fractions, each consisting of a small cluster of crowns, and replanting them in their new positions. The method is direct and economical and can be very effective, at least with some of the tougher and more tolerant plants. Many of the more attractive or interesting perennials do less well when treated in this brutal way, and serious losses occur, even with the most tolerant, during severe winters or in late cold springs when unestablished clusters of newly planted crowns are lifted from the ground by hard frosts, or beset by cold winds as their young shoots start to grow. This simple, casual method of propagation has become so well-established that most gardeners seem convinced that it provides the only means of propagating herbaceous perennials—it is not surprising that many gardens contain only a limited variety of these plants, consisting of those with so strong a hold on life that they are virtually indestructible.

Hardy, Tolerant Herbaceous Perennials that can be Divided during the Winter

Actaea	*Hesperis*	*Polygonum*
Aruncus	*Inula*	*Prunella*
Aster	*Lamiastrum*	*Rheum*
Centaurea	*Lamium*	*Saponaria*
Cerastium	*Lysimachia*	*Scabiosa*
Chrysanthemum	*Lythrum*	*Scrophularia*
Cimicifuga	*Melissa*	*Sedum*
Clerodendron	*Mentha*	*Senecio*
Eupatorium	*Oxalis*	*Solidago*
Geranium	*Phygelius*	*Stachys.*
Helenium	*Physalis*	
Helianthus	*Plantago*	

Propagation of Perennials in the Spring

The seasonal cycles of root growth described earlier (p. 18) provide the key to dividing perennials most economically and effectively, and suggest why propagating plants by division in the winter can be less than successful. This is a season when they are quiescent, and at a disadvantage if damaged in any way. In an inactive condition and at low temperatures plants can take a long time to make new root growth; their capacity to repair damaged tissues is low making them vulnerable to fungal and bacterial infections; the predatory attentions of slugs, millipedes and woodlice add to the destruction; and the rotting broken ends of roots and shoots provide soft starting points from which pathogens establish themselves. Conversely, plants in active growth, at seasons when temperatures are higher, are able to repair or replace damaged tissues rapidly, and quickly produce new roots.

Spring leading into early summer is one of the most favourable seasons, as the network of feeding roots is being produced. The leaves and shoots are growing rapidly at this time, and may have to be cut back severely while the roots establish themselves; that is a temporary setback, and very soon made good when the plants start to grow. Provided the divisions are set out into well-prepared soil, and not allowed to dry out if droughts occur, they very soon make new root growth quickly followed by renewed development of the shoots and flowers. Better results, but with more effort, may be obtained by potting up the divisions and protecting them in a cold frame for a short spell while they get started. But, at this season of the year, this is only likely to be justified in special circumstances. Perhaps when plants are being sub-divided into single crowns, to get the most from a small stock, or when dealing with rare plants, or when something is being propagated that is intolerant of disturbance and must be given special attention.

Perennials that can be Divided in the Spring

Aconitum	*Gunnera*	*Pontederia*
Artemisia	*Helictotrichon*	*Potentilla*
Arundinaria	*Helleborus*	*Primula*
Astilbe	*Helxine*	*Rodgersia*
Briza	*Macleaya*	*Salvia*
Caltha	*Meconopsis*	*Sidalcea*
Campanula	*Menyanthes*	*Stipa*
Carex	*Milium*	*Thalictrum*
Ceratostigma	*Miscanthus*	*Thymus*
Cortaderia	*Monarda*	*Trifolium*
Crepis	*Nymphaea*	*Tropaeolum*
Diascia	*Oenothera*	*Typha*
Dicentra	*Pennisetum*	*Veratrum*
Festuca	*Phalaris*	*Verbena*
Geum	*Phormium*	*Veronica.*

Propagation of Perennials in Late Summer

There is a third season when herbaceous perennials can be propagated by division, and one which until recently has been all but ignored by most gardeners. This is during the late summer in July and August, and it is a season which, in many ways, accords most closely with the plants' own growth cycles. This is the time of year when many of these plants produce the large roots which provide the foundation for all their growth and development during the following twelve months.

The method is very effective as a means of propagating the woodland species, a number of which are almost evergreen or semi-evergreen. After a midsummer period of quiescence shaded by the canopy of the leaves of the trees above them, these begin a period of very vigorous root renewal at this time, which is very obvious when plants like *Brunnera, Epimedium, Tellima, Geranium* and *Pulmonaria* are forked out of the ground and examined. All are in full leaf at this season, some are still in flower, and it goes against the grain to dig them up, and needs an act of faith to do so in order to take advantage of cycles of root growth which cannot be seen above ground level. Once the plants are dug up, provided the foliage is cut back to reduce transpiration, the plants can be confidently divided into single crowns. They will establish rapidly in the open ground if set out in nursery rows, well watered and protected by a shaded cloche for a couple of weeks while they get going again. They will prosper even better if potted up individually, and grown on in a cold greenhouse or a frame, with a little light shading until they are ready to be moved outside.

Divisions made at this season, potted up and grown on in a cold frame form well-developed, small plants by the autumn, and one of the great advantages of the method is that it provides a ready supply of young plants which can be planted at any time through the winter, and which become well-established by the time they start to grow in the spring.

Perennials that can be Divided During July and August

Achillea	Dicentra	Pachysandra
Aegopodium	Doronicum	Peonia
Ajuga	Echinops	Physostegia
Anaphalis	Epimedium	Primula
Anemone	Erigeron	Pulmonaria
Anthriscus	Filipendula	Pyrethrum
Aster	Fragaria	Rudbeckia
Astrantia	Gaillardia	Saxifraga
Bellis	Geranium	Symphytum
Bergenia	Hepatica	Tellima
Brunnera	Heuchera	Tradescantia
Campanula	Inula	Trollius
Centaurea	Liatris	Valeriana
Chelone	Ligularia	Viola.
Chrysanthemum	Omphalodes	
Coreopsis	Origanum	

CHAPTER NINE

Propagation of Shrubs, including Climbers and Roses

Introduction

Shrubs are long-lived, woody and persistent, producing a steadily expanding framework of branches on which to spread out their leaves and display their flowers. The persistent nature of the structures which make up a shrub, and other major differences from annual and herbaceous plants, fundamentally influence their survival strategies under natural conditions and affect their reproduction in very important ways. Seeds become less significant as a means of repeatedly replacing their parents, but play a very important part in the distribution of the species, and the ways that individuals find new locations in which to grow, and many shrubs produce fruits of one kind or another—succulent, sweet, brightly coloured and sometimes aromatic baubles, that are attractive to birds and mammals, which eat them, and then excrete the seeds within them far and wide over the countryside. The seeds produced by shrubs are very frequently subject to controls which delay their germination, and many possess robust and enduring seed coats which prolong the period they can survive in the ground before decay and disintegration exposes the embryo within them. As a result the seedlings of any particular vintage are likely to emerge over a period of many years, during which the slow processes of growth, decay and renewal in the surrounding vegetation may produce new locations in which seedlings can gain a roothold. Gardeners, impatient to produce plants, find these strategic delays irksome, and have devised methods of sowing and treating such seeds in ways which speed up their natural responses.

Shrubs also use the woody persistence of their branches as a means of self-perpetuation denied to most herbaceous perennials, whose shoots come and go with each successive spring and autumn. Bushes which lose their roothold may slump to the ground but their lower branches, pressed to the surface of the soil, are able to grow roots and produce new plants to create a thicket where only a single plant grew before. The branches of others bend to the ground as the stems develop and become top heavy; there they will root and give rise to a new plant which does the same in turn, to form a series of linear descendants leapfrogging away from a founder which, once upon a time, succeeded in establishing itself from a seed.

Layering may not yet be one of the forgotten arts, but amongst amateur gardeners it is certainly a neglected one. It is ideally suited to their needs, and provides a simple, reasonably effortless way to grow a few, large, well-rooted shrubs from those already in a garden; indeed, there is hardly any need to look after or even look at the plants while they develop.

Seed Germination

Shrubs that are Easy to Propagate

Many shrubs produce seeds which germinate with no difficulty, and do not require special treatments to persuade them to produce seedlings. These can be sown and treated in the same way as seeds of annuals (p. 56) described previously.

Shrubs that Germinate Easily

Abutilon	Eccremocarpus	Lavatera
Callistemon	Enkianthus	Leycestaria
Carpentaria	Eucryphia	Phlomis*
Celastrus	Exochorda	Piptanthus*
Chimonanthus	Fatsia	Pittosporum
Cistus*	Fremontodendron	Poncirus
Clethra	Fuchsia	Rhododendron
Colutea*	Genista*	Rhus
Coronilla*	Helichrysum	Salvia*
Cytisus*	Hydrangea	Schizophragma
Desfontainea	Indigofera*	Skimmia
Deutzia	Kalmia	Spartium*
Disanthus	Lavandula	Ulex.*

*These genera frequently produce seeds with hard, impermeable seed coats which prevent water reaching the embryos inside them. They will not germinate until the seed coat is destroyed or fractured by chipping or filing (p. 50).

Propagation of Shrubs that Require Multiple or Complex Conditioning Treatments

Many seeds of shrubs are very selective in the ways they control their germination. They may require closely defined conditions before they produce seedlings and, very often, these responses depend on several successive stages being completed. These may involve:

- The presence of immature embryos when the fruits ripen and the seeds are dispersed.

and/or

- The development of hard, impermeable seed coats, which must be broken down before water can reach the embryos.

and/or

- The presence of chemicals which inhibit seed germination, and must be diluted or leached away to remove their effects.

and/or

- Requirements for conditioning treatments at high or low temperatures, depending on the species involved, which enable the seeds to respond to favourable conditions later on.

The sequence of events can be complex, and may depend on critical timing of the different phases involved. A few days missed at the start of the processes can sometimes delay the emergence of seedlings by months, or even years. Conditions required vary from one species to another, and there may also be variations from one year to another in the extent to which particular seeds react to one or other of the phases involved.

Stratifying Shrub Seeds

One of the most satisfactory ways to cope with these problems is by a process called stratification. This geological term has been stolen and adapted reasonably enough to describe the procedure in which seeds (or fruits) and moist sand are combined in alternate layers. Unfortunately the word is also used, misleadingly and unnecessarily, to describe the technique referred to earlier (p. 88) in which seeds are exposed, after they have been sown, to conditioning treatments at low temperatures: this, less complex treatment is more accurately, and less ambiguously, referred to as chilling.

Nurseries producing trees and shrubs from seed have to handle large quantities of fruits, fleshy or otherwise, of all kinds, and these are not usually separated from the seeds within them. The small fruits are mixed in alternate layers with four or five times their volume of moist sand or grit, and packed into large pots or heavy-gauge plastic bags. These are buried in a pit of sand, or peat, or sawdust in a sheltered place outdoors, and left to weather for a while. Later, usually the following spring, or the one following, the pots or bags are disinterred, and the mixtures of grit, decayed remnants of fruits and the seeds are spread out to germinate in seed beds outdoors, or in seed trays under cover in a greenhouse.

Stratification resembles the chilling treatments used to condition seeds, and both are often called stratification. But there are substantial differences, the most obvious being the typically longer duration of stratifying treatments, which are designed to make use of a series of natural events during the autumn, winter and early spring, and quite often beyond, which destroy the remnants of the fleshy tissues of fruits, remove inhibitors, if they are present, break down hard seed coats and condition the embryos within the seeds.

The great advantage of this simple technique is that it uses natural conditions to expose the seeds to a succession of experiences, similar to those they would encounter in the wild, and provides a very good chance of an effective result. The disadvantages are that the sequence of experiences is dependent on the

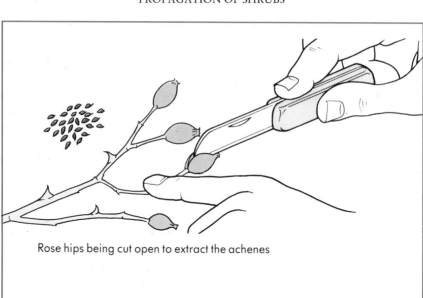

Rose hips being cut open to extract the achenes

Achenes mixed with moist grit and put into plastic bags

The bags with their mixtures of grit and achenes are packed into self-sealing plastic food boxes

The bags of grit and achenes are opened and their contents spread over the top of compost in pots

climate, and does not necessarily work well on seeds of plants which are naturally adapted to quite different climates. And, there is the difficulty of overcoming the natural preference of some seeds to germinate over a period of many years, rather than produce all their seedling together in a neat little batch.

The seeds of many shrubs are able to germinate as soon as conditions become favourable; more often their needs are more complex. They might have hard seed coats, containing an inhibitor, and enclosing embryos which produce seedlings, only after a two-month-long chilling period, when temperatures start to rise in the spring. An initial period at high temperatures would encourage the bacterial activity needed to break down the seed coats; and subsequently heavy rain or wet soils would remove the inhibitors. Only after these preliminaries were successfully completed would the embryo become responsive to the conditioning effects of a lengthy cold spell, and then at last be able to germinate when temperatures started to rise.

Variations on this theme apply to the germination of seeds of shrubs like *Berberis*, *Cotoneaster*, *Hippophae*, *Pyracantha*, *Rosa* and *Viburnum*. Different elements of the sequence vary in their importance and all are not always present, but the differences have much less practical significance than the resemblances. Timing becomes vital. If fruits and seeds are not available in time to start stratifying them early they miss the relatively warm weather during the autumn, and few or none will experience the degradation of their seed coats, which enables them to take up water or release inhibitors. As a result very few of the embryos will be able to respond to the cold conditioning treatment, and few seedlings will emerge during the first spring. The majority of the seeds will remain as they are, and pick up the natural seasonal sequence during the warmth of summer; respond to the following winter's chilling; and be ready to germinate when spring comes, nearly 18 months after they were originally stratified. A delay of perhaps only a few weeks in starting the stratification treatment can set back the time when seedlings appear by a year.

Fleshy fruits and seeds which are to be stratified should always be collected as soon as possible during late summer or early autumn, and the process set up without delay. If they cannot be obtained till later, for whatever reason, these seeds need not miss the preliminary warm spell which gets them going if the containers holding the stratified seeds are first put in a warm place for a month or six weeks, and then moved to more natural, colder conditions.

Amateur gardeners, for the most part, seek only small numbers of plants to satisfy their needs. This makes it practicable to remove the seeds from the fruits before stratifying them, and this is always worth doing when time allows. It also makes it possible to use small-scale equipment: by far the best containers in which to stratify seeds are polystyrene or polypropylene food containers. They come in all shapes and sizes, are practically air-tight, can be used repeatedly, and don't deteriorate even if seeds have to be kept in them in cold wet conditions for

Figure 9.1   Many shrubs produce seeds enclosed in fleshy fruits. Most of these must have conditioning treatments before they will germinate. Gardeners use a technique known as stratification to provide the conditions these seeds need to produce seedlings. The seeds are stratified by keeping them moist in plastic boxes for up to eighteen months. During this time the boxes can be kept, protected from the worst of the weather, in a cold shed, or a corner of the garden.

many months. They can be packed closely together, and if dropped they don't break.

Seeds are mixed with enough moist grit or vermiculite to make sure that each box is at least three-quarters filled. The boxes containing the mixtures can be packed together into a large wooden box, lined with heavy gauge polythene sheet, and left outdoors, or they can, more simply but not always so effectively, be put in an unheated garden shed on a convenient shelf. The containers must be opened from time to time, to look at their contents and find out whether any of the seeds have begun to germinate. During the winter this need be done only occasionally, but much more frequently when spring arrives. As soon as the first small roots can be seen emerging (and these are remarkably conspicuous) the containers should be removed, and their contents spread out in a layer about 2 cm ($\frac{3}{4}$ in) deep over potting compost in a plastic flower pot, or on a seed bed outside (p. 31).

The boxes containing seeds which show no signs of germinating are replaced and looked at again later, and can be kept for as long as is necessary. Sometimes it will be found that only a small proportion of the seeds in a particular container show the slightest signs of germinating. If so these can be removed and set out individually in a small pot, and the remainder left till another day. Treating seeds in this careful way provides a level of control which makes it possible to cope with erratic germination without having to lose seedlings which are not ready to emerge.

Growing on Seedlings

When the seeds do begin to germinate they will respond positively, enthusiastically even, to warm conditions, and amply repay being placed in a warm greenhouse. If possible the pots in which the stratified, chitted seeds are spread out should be placed immediately in the warm, and the seedlings given every possible encouragement to grow as fast as possible. Many shrubs, and trees, have remarkable capacities either to grow fast, or practically to stand still, during the early stages of their development. A good start early in the season, backed up later on by early transfer to individual pots, can gain as much as a year's growth compared with seedlings left to make the best of what nature provides in a cold frame or germinated out of doors on a nursery bed. Once well-established in individual pots young shrubs, growing strongly, should be ready to plant out into nursery beds by the middle of June where they can all but look after themselves.

Growing shrubs from seed successfully is very much a matter of saving time whenever possible. If they are treated appropriately, and if conditions favour rapid growth and development, they respond by germinating with few delays and growing away strongly. But, start the process a few weeks later or provide the growing plants with less than ideal conditions, and valuable time will be lost.

Shrubs that Frequently Depend on Stratification or Conditioning Treatments

Actinidia	*Aralia*	*Callicarpa*
Amelanchier	*Berberis*	*Caryopteris*

Chaenomeles	*Hypericum*	*Rhamnus*
Clematis	*Ilex*	*Rosa*
Cornus	*Ligustrum*	*Ruseus*
Cotinus	*Lonicera*	*Sambucus*
Cotoneaster	*Mahonia*	*Sarococcoca*
Daphne	*Nandina*	*Schisandra*
Decaisnea	*Parthenocissus*	*Stachyurus*
Elaeagnus	*Pernettya*	*Vaccinium*
Euonymus	*Pieris*	*Viburnum*
Gaultheria	*Prunus*	*Vitis.*
Halesia	*Pyracantha*	
Hippophae	*Raphiolepis*	

Propagation from Cuttings

Common sense, aided by experience, may be the best guide to choosing the right material to use as a cutting, and it is very difficult to define exactly what it is that makes one shoot likely to produce roots, but another, which may not be very dissimilar, a poor prospect. Yet, without being shown examples of what is, and what is not, suitable material it can be very hard to get it right.

It is sensible to avoid shoots which appear to be misshapen or diseased, although some willows are deliberately grown because they have a physiological disorder which causes their shoots to grow in an unusually thickened, and certainly misshapen, way, and this has no effect on the way their cuttings produce roots. Hardwood cuttings from roses may be taken from shoots which have lost all their leaves as a result of blackspot. They will produce roots perfectly well, and the plants raised from them will be neither more nor less susceptible to the disease than if they were taken from plants sprayed weekly to suppress all signs of the disorder.

Shoots which are short-jointed and compact are usually preferred to elongated ones, but there is little evidence to support this preference, and some which suggests that etiolated shoots which have been drawn up towards the light produce roots more easily than the compact products of high light intensities. Shoots which are particularly weak or spindly are usually avoided, as are those which are growing with exceptional vigour. There is more than a suspicion that gardeners choose shoots which display a normal, or average degree of vigour, not because these have been shown to be better, but because the middle way seems likely to be safer, and a nice batch of even cuttings, none too large nor any too thin, looks more satisfyingly efficient than a mixture of shapes and sizes.

Garden shrubs, grown for display, mature in such a way that they produce scarcely any shoots which can be used as cuttings, and the very short, twiggy growth typical of this condition is certainly most unpromising material. In nurseries shrubs grown as stock plants are pruned very hard year by year, and replaced frequently, to keep up a supply of the young shoots suitable for use as cuttings. This produces results which would be unacceptable in many gardens; nevertheless, anyone who intends to propagate some of the shrubs in the garden from cuttings is well-advised to take some thought for the morrow, if possible, so that when the time comes there will be shoots which can be used. Very often this can be done, satisfactorily and without spoiling the plant, by cutting back some of the branches very hard early in the spring, from which new growth will develop during the summer.

It is often taken for granted that there must be a best time to take cuttings of any particular kind of shrub, and that one of the keys to successful propagation would be to know when that is. Occasionally this is so, and the chances of propagating some plants do depend quite critically on doing things at the right season. More often than not a particular variety of shrub can be produced from cuttings with more or less equal ease, at different times of the year, using shoots at different stages of maturity. The method chosen depends on the convenience of the gardener, and on the equipment, facilities and time available.

Propagation from Immature Tip Cuttings

Many shrubs burst into growth with a flurry of activity in the spring. Supported by stored food reserves, their shoots grow quickly, and are extremely soft and tender; very susceptible to cold winds, or frost. But because they are growing rapidly, and their cells are dividing vigorously to support this growth, these shoots can often be used as cuttings, which produce roots much more readily than more mature shoots. There are some shrubs, lilacs (*Syringa* spp.) and the smoke bush (*Cotinus coggygria*) are two popular ones, which produce roots quite readily from cuttings of this kind, but are much more difficult to propagate from cuttings at any other season.

The tender tips of these young shoots require a little mollycoddling. They produce roots rapidly when set up in a gritty, well-drained cutting compost at a temperature of about 15°C (59°F), kept constantly moist (preferably under mist), and with as much natural light as possible. It is essential to encourage them to continue to function and grow actively so that they are able to produce roots very quickly. They will succeed under less than ideal conditions provided they are kept warm, and in an almost saturated atmosphere achieved by enclosing them in a frame covered with polythene sheet, or in an electrically heated propagator. These immature shoots have almost no resources within them, and can only support new growth by using sunlight as a source of energy. But, great care must be taken to ensure that high light intensities do not result in the cuttings being overheated in their enclosed containers. A compromise is necessary, by making sure they are sufficiently shaded to prevent their being scorched to death.

Figure 9.2   In the spring, during their first flush of growth, many shrubs can be propagated from cuttings made from the immature tips of their shoots. These cuttings produce roots and renew growth rapidly, but are very vulnerable and must have warmth and a constantly humid atmosphere around them.

(a) Tip cuttings of fuchsias

Tip cuttings dry out quickly, so use a plastic bag for protection as soon as they are removed from their parents

(b) Tip cuttings set close together in a gritty cutting compost

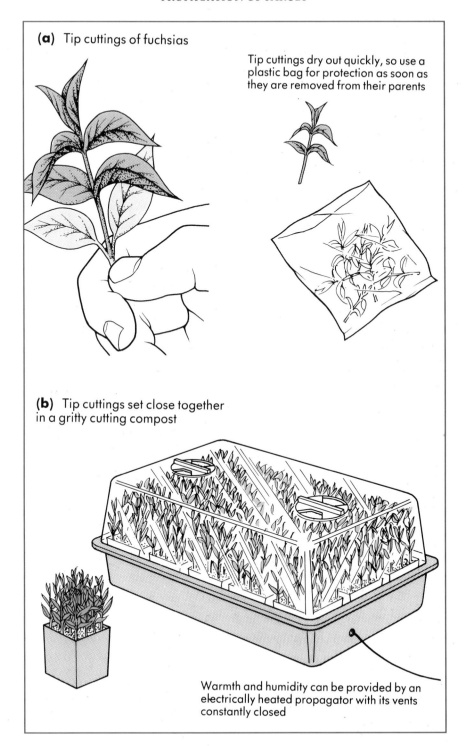

Warmth and humidity can be provided by an electrically heated propagator with its vents constantly closed

Nurseries make a practice of potting up stock plants and bringing them into a heated greenhouse during February to produce an early crop of cuttings. This is particularly worthwhile when growing Japanese maples and magnolias where well-grown cuttings must be produced if they are to survive the winter. Tip cuttings should be taken as nodal cuttings (p. 70), only 3–5 cm (1¼–2 in) long, as soon as the developing woody tissues inside the shoots can be felt at the point where the cut is made. These shoots dry out extremely easily once they have been cut off, and must be put straight into polythene bags as they are collected, and set up as cuttings under mist, or in enclosed containers, as quickly as possible. If all goes well, they will start to root very rapidly; lavenders, for example, can be well-rooted within a fortnight, and should then be given a liquid feed and potted up individually very soon after. These little cuttings should emerge as rapidly as possible from the propagator, to grow into well set up, bushy little plants by late summer.

Shrubs that can be Propagated from Immature Tip Cuttings

Abelia	*Drimys*	*Magnolia*
Abutilon	*Exochorda*	*Perovskia*
Acer	*Fabiana*	*Prunus*
Amelanchier	*Fuchsia*	*Salvia*
Callistemon	*Halesia*	*Syringa*
Caryopteris	*Helichrysum*	*Trachelospermum*
Ceratostigma	*Hoheria*	*Vaccinium*
Cestrum	*Kolkwitzia*	*Viburnum.*
Clematis	*Lavandula*	
Cotinus	*Lippia*	

Propagation in Summer from Semi-mature Cuttings

July and August should be the two great months for taking cuttings of shrubs. At this time most will have produced new shoots which have almost finished growing, but are still immature with their capacity for renewed regenerative growth still barely suppressed. The leaves on the shoots will be more or less fully grown, except for those at the tips, and the shoots themselves will be starting to firm up as the woody tissues within them develop towards maturity. At this stage they are much less tender than the soft tips produced during the spring flush of growth, and better able to tolerate conditions which are less than ideal. Ambient temperatures are also higher, and it is no longer essential to provide the cuttings with artificial warmth to encourage them to produce roots.

This is a season when gardeners can enjoy a bonanza, using very simple equipment, and gathering cuttings from an abundance of new growth. The shoots used can be cut off to produce either nodal, basal or heel cuttings usually between 5–10 cm (2–4 in) long, depending on the kind of shrub involved. They are relatively tough and tolerant, but must not be allowed to become dry and wilt, and it is good practice to put each batch into a polythene bag as it is collected, and not to leave them lying around untended for any longer than is absolutely necessary before setting them up.

The cuttings can be put in a shaded sun-frame, in a propagator, under polythene in a low tunnel, or under mist. The amount of care and attention they need is up to the gardener, but there are sacrifices to be made if the option chosen is for conditions where the cuttings are more or less left to look after themselves. Even this can be arranged. Cuttings can be set up in a propagator, or in a low, closed polythene tunnel (like a large cloche), and thoroughly watered-in. If they are then heavily shaded so that they receive only diffuse sunlight, they will remain moist, and the atmosphere around the leaves will remain saturated, for three or four weeks before they need be looked at again. Under these conditions they will produce roots, and establish themselves as young plants; not rapidly, but in time, and they will have needed so little attention that their caretaker's absence on holiday meantime would have passed unnoticed. If more attention can be given, they can be set up in conditions which expose them day by day to more of the warmth and light of the sun, and the more of each that they receive the faster they will produce roots. They will need more care, they dry out more rapidly, and in hot, sunny weather may need overhead sprays of water several times a day. If a mist unit is not available this must be done by hand.

Cuttings will produce roots, some within a week or two; some taking six to eight weeks depending on the conditions, and the kind of shrub they come from. Once roots have been formed and the young shoots start to grow again they should be drenched with a strong liquid feed (p. 82), weaned under slightly less protected conditions (p. 82), and then moved into a sheltered place, an airy greenhouse or cold frame is ideal, where they can continue to develop until the time comes to pot them up.

This marks a critical stage in their existence, and one where well-meant attempts to help them grow faster can have fatal consequences (p. 83). The problems arise from the time of the year that the cuttings are taken, the rate at which they form roots and the threat of winter looming ahead. Cuttings cannot usually be made from semi-mature shoots much before the beginning of July; late June at the earliest, but many can still be taken during August. Early cuttings, which produce roots rapidly with plenty of light and warmth from the sun, or from soil-heating cables, can be potted up individually during August, and will have plenty of time to establish themselves in their pots before winter. These are usually the chosen few. Cuttings taken later, or producing roots more slowly, particularly those kept for the sake of low maintenance in cool shaded conditions, will not be sufficiently developed to be potted up so early, and if potted later would not have time to establish themselves before growth ceased in the autumn. As a simple rule of thumb, all cuttings which are not well and truly ready to be potted up by 1st September are better left as they are till they start to grow again in the spring. Almost all cuttings are likely to be more at risk when overwintered individually in pots, than when left undisturbed in the containers in which they were rooted; even though they may appear to be uncomfortably overcrowded.

Shrubs that can be Propagated from Semi-mature Cuttings

Abelia	Fatshedera	Phyllodoce
Abutilon	Forsythia	Physocarpus
Actinidia	Fothergilla	Pieris
Akebia	Fuchsia	Piptanthus
Azara	Gaultheria	Pittosporum
Ballota	Genista	Polygonum
Buddleia	Grevillea	Potentilla
Bupleurum	Griselinia	Prunus
Callicarpa	Halimiocistus	Punica
Callistemon	Halimium	Pyracantha
Calycanthus	Hebe	Rhamnus
Camellia	Hedera	Rhododendron
Caryopteris	Helianthemum	Rhus
Ceanothus	Helichrysum	Ribes
Celastrus	Hibiscus	Rosa
Ceratostigma	Hoheria	Rosmarinus
Cestrum	Holodiscus	Ruta
Chaenomeles	Hydrangea	Salvia
Choisya	Hypericum	Sambucus
Cistus	Hyssopus	Santolina
Clematis	Ilex	Sarcococca
Clethra	Indigofera	Satureia
Colutea	Jasminum	Schisandra
Convolvulus	Kerria	Schizophragma
Coronilla	Kolkwitzia	Senecio
Corylopsis	Lavatera	Skimmia
Cotoneaster	Leptospermum	Sorbaria
Crinodendron	Ligustrum	Spiraea
Cytisus	Lonicera	Stachyurus
Daphne	Muelenbeckia	Symphoricarpos
Desfontainea	Myrtus	Symplocos
Deutzia	Nandina	Syringa
Diervillea	Osmanthus	Teucrium
Dipelta	Osmarea	Trachelospermum
Disanthus	Parahebe	Ulex
Elaeagnus	Parthenocissus	Viburnum
Enkianthus	Passiflora	Vinca
Escallonia	Pernettya	Vitis
Eucryphia	Perovskia	Weigela
Euonymus	Philadelphus	Wisteria.
Exochorda	Phlomis	
Fabiana	Photinia	

Propagation from Hardwood Cuttings of Deciduous Shrubs

As the growing season comes to an end and autumn turns into winter the leaves fall from deciduous shrubs; their shoots become firm, hardened by the woody tissues within them; and their cells fill with starch, and other storage reserves, set aside to enable them to survive the winter and grow away strongly the following spring. These tough sticks may not appear to be promising material to use as cuttings: but, they are better than they look. They may not be growing actively, but they are all prepared to do so the following spring and their tissues contain the regenerative cells which will divide to get growth going again when it comes. They are extremely tolerant of poor conditions, and forgiving of treatment which would lead to failures with other less completely developed shoots, and they are very well able to support themselves from the reserves packed away inside them; a self-sufficiency which enables them to survive and prosper with few demands on the gardener's time or attention.

These hardwood cuttings provide one of the easiest means of propagation for hard-pressed gardeners, even for those who are short of time and unconfident of their skills. Many are able to produce roots successfully in a sheltered bed in the open ground; others benefit from the protection of a cold frame, and all will do better with this protection when winters are severe. The best shoots to use are vigorous, straight young shoots, of the kind which appear spontaneously after plants have been cut back hard. These are produced during the normal course of cultivation, with bush roses that are pruned hard each spring and with shrubs like *Buddleia* which are cut back in a similar way to keep them in shape and encourage abundant flowers in late summer. Sometimes, when the need can be anticipated, it may be worth cutting other shrubs back likewise, although this is not their normal treatment, to make sure of a supply of cuttings.

Cuttings of this kind have, by long tradition, been taken throughout the winter from leafless shoots, and it has often been looked on as a convenient job to occupy the dull months of December and January. The shoots used look and feel just the same whether they are removed in November, February or March, but although unchanged to outward appearances their internal arrangements change as time goes by, and these changes affect the way that roots are formed and, in particular, the proportion of shoots which produce roots successfully. Unfortunately for our convenience, roots are least likely to be produced from cuttings made in the dead of winter, and should be done as the leaves fall during late October or within a month of that event. If the cuttings are not taken then it is better to delay their removal until March, only a short while before they come into growth again.

These cuttings are prepared from shoots which are fully mature with very firm wood and, unlike all the other cuttings described previously, the best parts to use are not the tips. Long lengths of shoot can be cut from the plant, provided all the growth was made during the preceding summer, and these can be sliced into sections each with about four buds, and measuring something like 10–20 cm (4–8 in) long. The cuttings are stuck into cutting compost (p. 76) in boxes or large pots in a cold frame, or into a vee-shaped slit in the ground filled with cutting compost. It is not necessary to bury them deeply, nor for them to project far out of the ground. Ideally their uppermost buds should be just above ground level, with the bud below only a short distance beneath the surface.

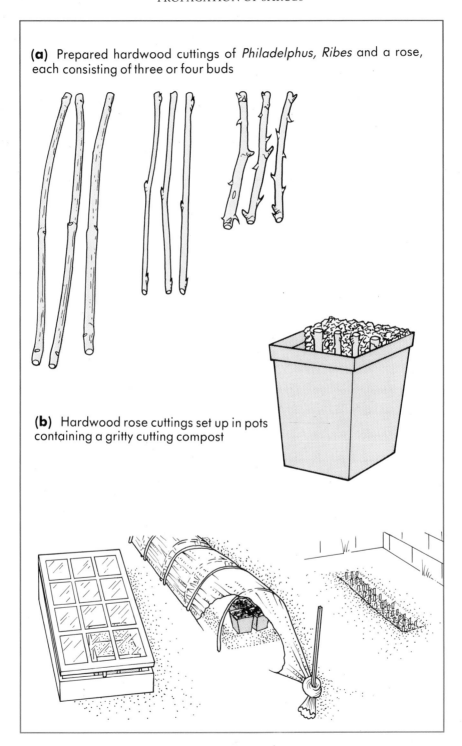

(a) Prepared hardwood cuttings of *Philadelphus, Ribes* and a rose, each consisting of three or four buds

(b) Hardwood rose cuttings set up in pots containing a gritty cutting compost

Once in place very little more need be done. Cuttings set up in pots or boxes can be potted up individually once their roots are well-developed, usually during the following spring, and grown on under cover in a greenhouse, or in a sheltered place outside. These should grow into well-established plants by autumn, large enough to plant into their permanent positions in the garden. Cuttings which are expected to produce roots under more exposed conditions outside will progress more slowly and it is better to leave these, and the plants that develop from them, where they are until the following autumn or winter, when they can be dug up and replanted, either in their places in the garden, or lined out in a nursery bed.

Deciduous Shrubs that can be Propagated from Hardwood Cuttings

Abutilon	*Forsythia*	*Ribes*
Berberis	*Genista*	*Rosa*
Buddleia	*Hypericum*	*Rubus*
Caryopteris	*Jasminum*	*Salix*
Ceanothus	*Kerria*	*Sambucus*
Cornus	*Lavatera*	*Spiraea*
Coronilla	*Ligustrum*	*Stephanandra*
Cotinus	*Lonicera*	*Symphoricarpos*
Cotoneaster	*Philadelphus*	*Tamarix*
Deutzia	*Photinia*	*Viburnum*
Diervillea	*Physocarpus*	*Vitis*
Dipelta	*Polygonum*	*Weigela.*
Elaeagnus	*Potentilla*	
Euonymus	*Prunus*	

Propagation from Hardwood Cuttings of Broad-leaved Evergreens

Shrubs which retain their leaves throughout the winter do so because they come from parts of the world where the winter climate is not insupportably severe. Most hardly grow during this period, and their leaves are likely to be thick and robust as a protection against storms and cold, but they are able to photosynthesise, and support the development of new organs, including roots, when the need arises.

That need arises as soon as a shoot is removed to be used as a cutting. These are very similar to the hardwood cuttings of deciduous species already described, but the presence of leaves, even such tough leaves, makes them more

Figure 9.3   The leafless shoots of a great many shrubs, and some trees, can be used to provide cuttings during the winter. These hardwood cuttings can be used to produce plants successfully, with much less care and attention than the more vulnerable, less self-sufficient cuttings made from leafy shoots. They can be set up in pots protected by a coldframe or cloche, or inserted in a slit cut in the open ground, in a sheltered corner of the garden, and filled with sand or grit.

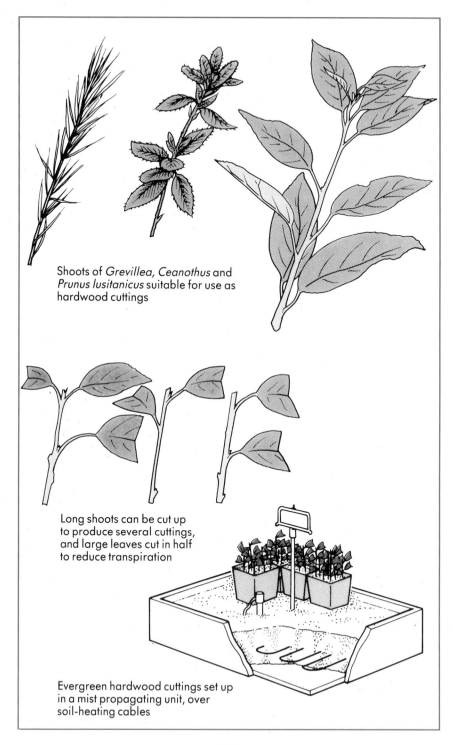

Shoots of *Grevillea, Ceanothus* and *Prunus lusitanicus* suitable for use as hardwood cuttings

Long shoots can be cut up to produce several cuttings, and large leaves cut in half to reduce transpiration

Evergreen hardwood cuttings set up in a mist propagating unit, over soil-heating cables susceptible to desiccation and damage, and they require some protection to avoid this. It also makes it possible for the cuttings to respond more positively to artificial light and extra care. Broad-leaved evergreens can be propagated from cuttings taken in the late autumn and given the protection of a cold greenhouse or cold frame through the winter. When this is done a careful balance must be maintained between letting them become too dry, when they shrivel, or too damp, when they are liable to rot. If facilities are available it is better to take the cuttings in February or March. The advantages and disadvantages are evenly balanced: taking cuttings before the winter increases demands on good management, but provides an insurance against losses of plants which are not totally hardy; taking them in early spring, provides for easier management but depends on more expensive facilities, and, if the winter has been a hard one, some of the parent plants may already be dead, destroyed by frost or cold winds.

The cuttings taken early in the spring are fully prepared to break out from their winter quiet into very active growth, and all they need are the right conditions: a minimum temperature of 15°C (59°F) in the compost around their bases—if the tops are in cooler conditions so much the better—and constant humidity using light mist or hand spraying. Cuttings taken at this time come into growth immediately, they are able to produce roots rapidly and are much less susceptible to decay than the slowly moving shoots removed in the autumn. The young rooted cuttings should be potted up as soon as possible and grown on through the summer to make substantial plants by the autumn. Good growing conditions with artificial warmth to start with, and the protection of a greenhouse later, will be amply repaid.

Evergreen Shrubs that can be Propagated from Hardwood Cuttings

Arbutus	*Euonymus*	*Phlomis*
Aucuba	*Fremontodendron*	*Photinia*
Azara	*Garrya*	*Raphiolepis*
Berberis	*Grevillea*	*Rhamnus*
Buxus	*Griselinia*	*Rhododendron*
Carpentaria	*Hedera*	*Rosmarinus*
Ceanothus	*Ilex*	*Ruta*
Choisya	*Itea*	*Salvia*
Cistus	*Kalmia*	*Sarcococca*
Corokia	*Laurus*	*Senecio*
Cotoneaster	*Lavandula*	*Skimmia*
Desfontainea	*Leptospermum*	*Vaccinium*
Drimys	*Ligustrum*	*Viburnum*
Elaeagnus	*Lonicera*	*Vinca.*
Escallonia	*Myrtus*	
Euchryphia	*Olearia*	

Figure 9.4   Many broad-leaved evergreen shrubs can be propagated during the winter from cuttings made from mature shoots produced during the previous summer.

Propagation from Single Bud Cuttings

Most of the familiar cuttings prepared from shoots consist of lengths of stem each bearing a number of buds. Normally they are cut off, either at the point where side branches emerge from a larger stem, or at the nodes. Many climbing plants produce lengthy, attenuated stems with a considerable distance between each node, and cuttings prepared in the normal way from plants like clematis or vines are extremely awkward to handle, and to set up in pots of cutting compost. Even the shortest can be 30 cm (1 ft) long or more, and they become entangled with one another, they flop about, and more often than not while putting one in, two more are dragged right out of the compost. One way to overcome these practical problems, is to cut these long stems into short sections by a series of cuts made through each internode.

The cuttings are formed by abandoning one of the golden rules of propagation: to make all cuts immediately beneath a node. Instead the stems are cut through 2 or 3 cm ($\frac{3}{4}$–$1\frac{1}{4}$ in) beneath the points where leaves join the stems at the nodes, and then a second cut is made immediately above each node. This produces a neat little cutting, consisting of no more than one pair of leaves with a convenient piece of stem beneath them to provide a peg to stick into the compost. These cuttings are then treated like any other: clematis taken between June and August are the equivalent of semi-mature cuttings (p. 130); vines propagated from their bare rods in early winter will respond to the same conditions as hardwood cuttings.

It may seem too good to be true, but plants which produce cuttings that are awkward to handle in conventional ways just happen to be the ones which oblige us when we break the rules. And, in fact, plants which can be propagated from internodal cuttings containing a single bud are by no means confined to long-stemmed climbers. Many other shrubs can be propagated in the same way, but in most cases traditional cuttings are plentifully available, easier to handle and produce a plant more quickly. Single bud cuttings are a way of getting the most from a few shoots, since every bud is a potential plant. This advantage may be offset by the time taken to produce a plant from a single bud which is usually longer than with other, larger types of cutting, but it provides a useful but neglected way to produce plants in certain circumstances. Apart from propagating climbers, its use is mostly confined to one or two particular situations, where other methods of propagation have drawbacks of one kind or another.

Roses and Mahonias are amongst the plants which gardeners sometimes find worth propagating in this way. Roses are most frequently propagated commercially by budding, which is a very economical method of using regenerative tissue to produce large numbers of plants, but may be a daunting prospect to those who lack the skills to do it. Most roses can also be propagated from

Figure 9.5   Cuttings of climbing plants and vines with long, trailing stems become inconveniently long and difficult to handle, when prepared in the usual way. These are much easier to handle when prepared from small cuttings each containing a single bud or pair of buds. These clematis cuttings have been prepared from a long trailing stem. Very often these stems hang upside down, and it is important to note that the true position of the buds is in the angle above their accompanying leaf. It is usually convenient to remove all but the first pair of leaflets on each leaf.

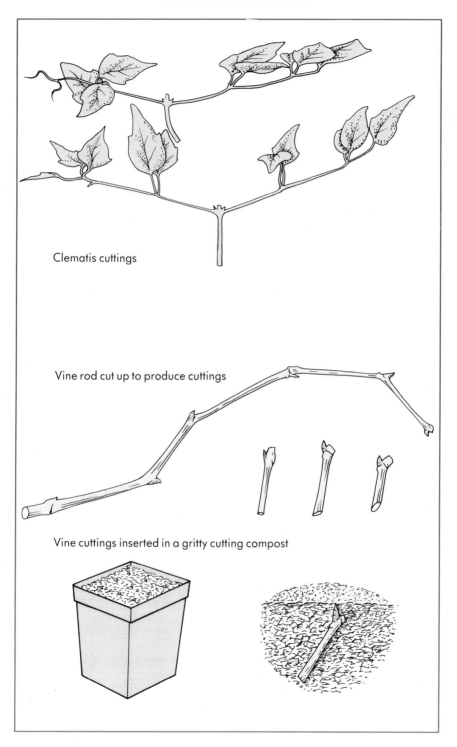

Clematis cuttings

Vine rod cut up to produce cuttings

Vine cuttings inserted in a gritty cutting compost

hardwood cuttings (p. 135), but, since each cutting is made up of at least four buds, and not all cuttings by any means produce a plant, this is a more extravagant use of resources, especially when there is a need to produce large numbers of plants from a few chosen individuals. So, single bud cuttings of rose stems is an economical way to make the most of what is available. The best time to do this is during late July, after they have completed their first flowering, or in the case of many shrub roses, finished flowering, and when they have produced strong young shoots which can be removed and sliced up from tip to toe to provide numerous small cuttings. These are then best set up close together in pots or small trays and put into a covered frame or propagator in essentially similar conditions to those used for semi-mature cuttings of other shrubs (p. 130).

Propagation of Silver-leaved, Aromatic and Semi-tender Shrubs in the Autumn

In our gardens, we grow many shrubs which come from areas with Mediterranean climates or from cool temperate parts of the world, such as New Zealand, where winters are not very cold. Many of these survive average winters in gardens in other colder, temperate regions but are killed when conditions are very severe. Almost all are evergreen, and almost all continue to grow well into the winter or even right through it, without becoming quiescent. It is a sensible precaution to take cuttings of these shrubs in the autumn to provide replacements in case they are killed, and this is a very good season to do so for other reasons. The plants are able to provide plenty of material in the shape of young shoots produced during the previous summer; they produce roots readily at this time of year; and, if provided with a very little protection and warmth through the winter, will be well-established by the time spring comes.

A reassuringly simple point to remember is that these cuttings are tough. By September their shoots have become fully mature, and are filling up with the storage reserves which will see them through the winter, and boost their growth in the spring. Their leaves are fully formed with thick cuticles and other features which enable them to survive even if they are a little short of water.

The shoots to use should be fully mature, nodal or heel cuttings about 5–10 cm (2–4 in) long. These can be set up in containers in a normal cutting compost (p. 76), using one which is gritty and drains impeccably. These cuttings are tolerant of some exposure, and do not need close confinement in a humid atmosphere; they will respond well if set up on a bench in a well-ventilated greenhouse, protected by the glass overhead. If possible, soil-heating cables should be used to provide them with a temperature of about 15°C (59°F) during the first month or six weeks when they are producing roots. During the rest of the winter the temperature should be reduced; all that is necessary is to keep the cuttings frost-free but no more. Throughout the time when they are producing roots, and right through the winter, watering must be very strictly controlled. Damp will kill these cuttings easily; drought, unless extreme, will scarcely

Figure 9.6   Many roses can be propagated from single bud cuttings during July and August.

Young rose shoot removed
during July or August

Sections of stem containing
one bud and a leaf

Cuttings in 9 cm square plastic
pots; set up in a propagator
or frame

affect them. In the spring they should be given a liquid feed during February and potted up individually as soon as they show signs of renewed vigorous growth.

Semi-tender, Cool Temperate Shrubs Propagated in the Autumn

Abutilon	*Fabiana*	*Passiflora*
Achillea	*Fuchsia*	*Penstemon*
Azara	*Grevillea*	*Phlomis*
Ballota	*Griselinia*	*Pittosporum*
Buddleia	*Hebe*	*Punica*
Caryopteris	*Helichrysum*	*Rosmarinus*
Ceanothus	*Hypericum*	*Ruta*
Cestrum	*Laurus*	*Salvia*
Cistus	*Lavandula*	*Santolina*
Convolvulus	*Lavatera*	*Senecio*
Corokia	*Lippia*	*Teucrium*
Coronilla	*Melianthus*	*Zauschneria.*
Escallonia	*Myrtus*	
Eupatorium	*Olearia*	

Propagation from Root Cuttings

A number of shrubs can be propagated from root cuttings. The roots of several, including roses, have a predisposition to form shoots, which is stimulated by cutting them into sections. They should be treated in the same way as root cuttings of herbaceous perennials (p. 111) making allowance where necessary for their greater size. Usually it is unnecessary to dig up a bush entirely; a hole dug beside it is enough to find a root or two which can be removed and used to produce a few cuttings.

Shrubs that can be Propagated from Root Cuttings

Aralia	*Chaenomeles*	*Rhus*
Campsis	*Clerodendron*	*Rosa.*
Celastrus	*Embothrium*	

Division

Reproduction from Suckers

Numerous shrubs, including many roses, produce suckers naturally as a simple and effective method of forming a thicket of stems progressing outwards from a well-established base. By and large gardeners regard suckers rather unkindly, associating them with the problems they experience with budded and grafted plants of roses, lilacs, viburnums, plums and cherries, and view them as a nuisance. The strong shoots from grafted rootstocks which compete with and may eventually suppress the varieties worked on them deserve this reputation, and these suckers are, indeed, of little or no use to most gardeners.

Suckers produced by plants growing on their own roots are a different matter, and should be assessed for the opportunities they offer. They provide a useful, ready-made source of plants ideally suited to the needs of amateur gardeners. All that need be done is to dig them up with a portion of root and plant them elsewhere. Some with very few fine feeding roots, will need to be potted up and nursed a little while they establish themselves, but even these provide a rapid, and easy supply of well-grown shrubs.

Shrubs that can be Propagated from Suckers

Amelanchier	Pachysandra	Sorbaria
Ceratostigma	Passiflora	Spiraea
Chaenomeles	Pernettya	Stephanandra
Clerodendron	Perowskia	Symphoricarpos
Corylus	Phyllodoce	Syringa
Diervillea	Prunus	Vaccinium
Elaeagnus	Rhus	Yucca.
Gaultheria	Rosa	
Kerria	Sarcococca	

Propagation from Layers

A great many shrubs produce trailing branches which develop roots naturally at places where they touch the ground. Others layer themselves when their branches are weighed down by drifts of snow or are partially broken by storms so that they reach the ground. In either case the formation of roots becomes even more likely when the grounded stems or broken branches become lightly covered by soil or leaf litter.

These natural processes can easily be imitated and used as a means of propagation known as layering; the individual rooted stems being called layers. On nurseries this is usually done in an organised and relatively complex way, which involves preliminary preparation of the parent plants, and the soil around them, to make sure that the greatest possible number of layers are produced. On a smaller, more domestic, scale it can be done quite casually, with a branch or two of a shrub growing in the garden in such a way that it scarcely shows, and simple forms of layering of this kind are an ideal way to produce a few plants.

Layers can be produced using quite large side branches bent down to the ground and held firmly in place while they produce roots. However, young whippy stems are more likely to form roots, and will do so more quickly, than larger and more mature branches, and if not already present can often be encouraged by cutting a few of the more mature branches back almost to ground level in the spring. Their stumps should produce strong shoots during the summer which can be layered the following autumn. As a rule, layered branches produce roots more rapidly when their stems are damaged close to the point where they meet the ground. This can be done in various ways; the usual recommendations being to twist the stem as it is bent round, or to make a slanting cut along its underside so that a tongue is produced which opens as the further end of the stem is bent up. Both these are effective methods, but it is even simpler to use a sharp knife to remove a sliver of bark on the underside of the branch at the place where it will be pressed down in close contact with the ground. The stems are kept firmly in place while they produce roots by pegging them down, with wooden pegs cut from forked branches. The end of the layer, beyond the point where it meets the ground, is the part that will eventually produce a new plant, and this should be trained by tying it to a stout bamboo cane so that it rises up into the air almost vertically. The bamboo must be firmly pushed deep into the ground and come about two-thirds of the way up the stem of the layer. Bamboo canes which project beyond their layers become booby traps which can blind when eyes peer through the foliage to see whether roots are being formed.

Once the layer is firmly pegged down and tied in position its lowest point can be covered with a shovelful of a 50:50 mixture of grit and garden soil and then little more need be done except wait for it to produce roots. This can be a lengthy process, especially under the rather haphazard conditions with little previous preparation that are normally inevitable in a garden. It may take nine months, it may take as long as two years and the best time to search to see if there are any roots is during the winter, by scrabbling in the soil at the base of the shoot. When roots are discovered it is natural enough to want to press on and cut the rooted layer free, dig it up, and plant it where it can take its place in the garden. Usually this works and the young plant establishes itself satisfactorily. If time and circumstances allow, and patience is not quite exhausted, it is preferable to do the operation in two stages; cutting the layer free, but leaving it in position undisturbed, and then digging it up, and replanting it a few months later after a short period of adjustment to independence.

Layering may appear to be a lengthy process and one that demands exceptional patience, and perhaps this is the main reason why so few people attempt to grow shrubs from layers. But, although first appearances may seem discouraging, layers are a very useful way of obtaining plants. They can be prepared initially from much larger pieces of plant than any cutting could be, and they continue to grow and develop throughout the time when they are producing roots. Consequently, by the time a layer is independent and ready to plant in the garden it should have developed into a substantial plant which compares more than favourably with shrubs produced from cuttings over the same length of time. Throughout this period it has required practically no attention at all—producing a ready-made plant, which just has to be dug up and replanted; even the most labour-saving methods of producing plants from cuttings require much more careful attention than that.

Figure 9.7   Many shrubs can be propagated by bending quite large branches to the ground, and pinning them down until they produce roots. This process may take a year, or even eighteen months, but it can be a very economical and simple way to produce a few large replacement shrubs for a garden.

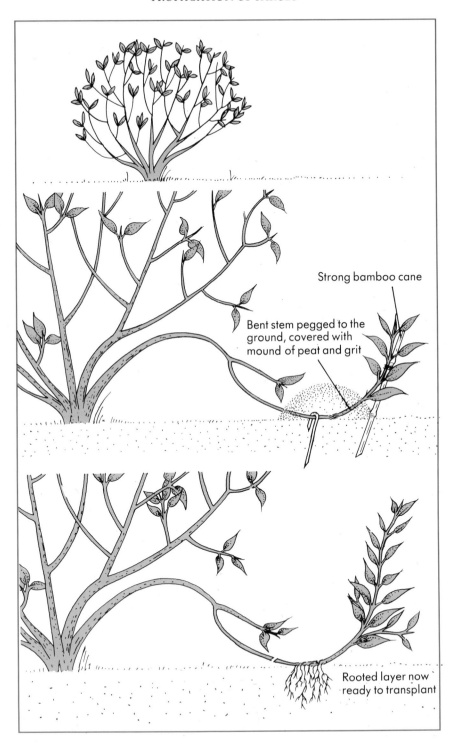

Strong bamboo cane

Bent stem pegged to the
ground, covered with
mound of peat and grit

Rooted layer now
ready to transplant

Shrubs that can be Produced by Layering their Branches

Actinidia	Elaeagnus	Phyllodoce
Akebia	Enkianthus	Pieris
Amelanchier	Eucryphia	Rhododendron
Calycanthus	Euonymus	Rhus
Carpentaria	Fothergilla	Salix
Celastrus	Gaultheria	Schisandra
Cercidiphyllum	Halesia	Schizophragma
Chimonanthus	Hedera	Stachyurus
Clematis	Ilex	Stephanandra
Clethra	Kalmia	Syringa
Cornus	Laurus	Trachelospermum
Corylopsis	Ligustrum	Vaccinium
Corylus	Lonicera	Viburnum
Cotinus	Magnolia	Vitis
Daphne	Osmanthus	Wisteria.
Disanthus	Parthenocissus	
Drimys	Pernettya	

Propagating by Air-layering

Shrubs can be layered to the ground to produce new plants with very high hopes of success. But some grow with such upright branches that it is almost impossible to bring their shoots down to the ground. A technique known as air-layering provides a method of bringing the ground to the branch. The principle is simple enough; in practice it is not quite so easy.

In principle, and, in brief: young strong-growing, straight shoots are selected, or are encouraged to develop by cutting one or two branches back hard. Moist, water-holding material, and sphagnum moss is most often used, is then wrapped around sections of stem, including a node, and enclosed in layers of opaque waterproof plastic sheet. The top and bottom of the enfolding wraps are securely sealed with tape to prevent water running in down the stem from above or evaporating from the moss, and the package is then left alone while the enclosed sections of stem produce roots, which make their way into the surrounding moss. Sometimes, attempts are made to encourage roots to form by wounding the stem on one side by slicing off a sliver of bark, or the stem may be girdled by removing a narrow ring of bark immediately below the point where, it is hoped, roots will be produced.

Air-layering is used to propagate a number of house plants, which grow rapidly in warmth and shelter, and, under these conditions, it is an effective and regularly used means of producing new plants. Problems arise when similar methods are used on hardy plants growing out of doors, where they are exposed to every kind of weather. Roots are produced much more slowly and it is difficult to devise a weatherproof package which keeps water out yet remains moist throughout the period when roots are being formed, until they emerge and establish themselves in the sphagnum moss. Seals formed by taping the polythene sheet usually fail to remain watertight when exposed to extremes of heat or cold. Rainwater runs down the stem and, if it penetrates the packaging, can saturate the moss with water, causing waterlogging at any time and damage from freezing during cold weather. Precautions must be taken when making up the wrapping to ensure that it is opaque, because light inhibits root formation, and that it is well enough insulated to prevent the interior becoming lethally hot when the sun shines directly on to it.

If these problems are solved and roots are produced as planned, the rest is relatively simple. The shoots are cut off during the winter, just below the area where roots have been produced, and potted up individually to be grown on, preferably in a greenhouse, till the following autumn. They can then be lined out in a nursery bed, and eventually planted in the garden. The description of the problems involved in producing plants of hardy shrubs, and trees, by this method may sound discouraging, and serves to emphasise the point that air-layering is not usually worth trying if other, less technically awkward methods, can be used. However, there are a number of trees and shrubs which are difficult to propagate by any means, and inevitably these tend to be rare and choice, for which this method does offer a gleam of hope. The fact that it is achieved against the odds makes success all the sweeter when the result is a triumph.

Shrubs that can be Propagated from Air-layers

Chimonanthus	*Eucryphia*	*Peonia*
Cornus	*Fatshedera*	*Pieris*
Cotinus	*Fatsia*	*Piptanthus*
Enkianthus	*Magnolia*	*Rhododendron.*

Propagation By Tip-layering

Propagation by tip-layering is almost confined to brambles, of one kind or another, which practice it naturally as their normal means of reproduction. It can be adapted for use with other shrubs which have long trailing stems, but few lend themselves to it so naturally, and most are easily produced by other methods. In contrast to air-layering it is extremely easy to do successfully.

The long, arching shoots of different species of *Rubus* are bent over, and their tips pegged down to the ground or into pots of potting compost during July and August. A simple way to ensure that they stay securely lodged in their pots is to thread the tip of the shoot through one of the drainage holes in the bottom, and bring it up through the pot before embedding it in a mixture of grit and garden soil. These tips rapidly produce roots, and their buds start to grow up, up and away to form a new ascending shoot. During the winter they can be cut free from their parent plants and set out in their places in the garden. If very small, and many will be, their position should be marked with a cane, but they will develop so rapidly the following summer that they are perfectly capable of holding their own.

CHAPTER TEN

Propagation of Trees

Introduction

Trees are large and long-lived, and the processes by which their seeds are distributed, develop into established seedlings, which grow into saplings and eventually mature, can appear unhurried, dilatory even, to a gardener impatient to see results. Seeds play a decisive part in the ways trees survive, but over an immensely longer time scale than other plants with shorter life spans. Seedlings as they emerge face a long future of changing conditions all around them, while they grow, and the surrounding vegetation fills out or declines; for many years they will be among the smaller more vulnerable elements in this pattern of survival and decay. Only a minute number do survive, and those that do are the ones which find themselves in positions where fate provides them with a chance, and which are able to hang on and use that chance when it comes.

In consequence many trees produce seedlings which are able to persist under adverse conditions, making little growth, but slowly developing their roots beneath the ground, and building up their reserves above it, so that when an opportunity for more rapid development arrives they are able to grow very quickly to take advantage of it. These alternative options for stasis or dynamic growth are very significant in the garden. Trees which are planted in places where opportunities for development are limited can spend years doing remarkably little: the ultimate expressions of this ability are found in the art of Bonsai. With less skilful care this lack of opportunities becomes neglect, and the kindest thing that can be said about some of the twisted, stunted trees which result is that they look 'characterful'. On the other hand, given progressive care, plenty of light, nutrients, water and warmth, trees can grow extremely rapidly. Seedling birch trees, germinating in the spring and left in a seed bed in the open through the summer may be no more than 20 cm (8 in) high when their leaves fall in the autumn. Their siblings, lifted and potted up during April, grown on in a greenhouse and repotted early in July may be close to 2 m (6½ ft) by the end of the season; nearly ten times the height.

Nurseries usually grow trees in very large numbers. Amateur gardeners may wish to grow a wide variety, and their seeds provide a convenient way to find unusual kinds but, unless they are planning to plant forests, they usually need only a very few of each. The key to success is simple; it is to be ruthless in limiting the numbers of each to no more than the bare minimum actually needed, and very often that is only one, two perhaps to allow for losses; and to look after the ones that are grown really well. Hardy trees might appear to be almost the last plants one would choose to occupy precious space in a greenhouse during the summer; but, if they are looked after, and each one given the extra care and attention which will produce a well-grown specimen tree, few other plants occupy the space more profitably.

Seed Germination

Sowing Seeds of Trees and Growing On Seedlings

The seeds of many species of trees need a conditioning treatment at low temperatures (p. 88) or a period of stratification (p. 123) before they will produce seedlings. Most frequently the former is sufficient but hollies and many of the thorns have complex requirements involving a sequence of events, which can only be satisfied by lengthy stratifying treatments. If possible most tree species with small, dry seeds should be sown during November and December to provide time for chilling treatments to be effective. The larger nuts are very likely to be eaten if sown during mid-winter; these are treasure trove to birds, rodents and squirrels, and easy to find and hard to protect. Most can be pre-germinated in polythene bags of moist peat and set out by hand in late February or during March to reduce the length of time they are at risk.

Nuts of all kinds will only remain in good condition (p. 180) if they are kept cool and moist. These should be mixed with peat or moss in plastic bags from the moment they are collected until they are sown; even a few days in a dry room fatally undermines the vigour of fragile kinds like acorns and chestnuts. During February or March they can be sown in an outdoor seed bed if numbers are wanted, or in plastic pots in a frame or a greenhouse when a few are enough. Seeds sown in the greenhouse will germinate rapidly; they can be potted up individually and grown on, even, if possible, with a little artificial warmth until the end of May, when they can be moved into a larger pot. If kept in the greenhouse or in a plastic tunnel and cared for through the summer they can make large plants by the autumn, when they should be stood outside to lose their leaves naturally, and harden off before winter comes. This accelerated growth is well worth going for when small numbers of trees are wanted to furnish a garden quickly.

Under normal commercial conditions much larger numbers of seedling trees are grown more slowly but steadily; these methods can be scaled down, and are very useful when moderate numbers of trees are needed, but time is not pressing. The seeds are sown on to seed beds set up in a sheltered place outside. These are first lightly forked and raked, and then covered with a 2–3 cm ($\frac{3}{4}$–$1\frac{1}{4}$ in) layer of grit. Small seeds like birch are scattered on the surface and raked in so that the majority lie just below the surface; medium-sized seeds including the pagoda tree and hornbeam should be sown on the surface and lightly covered with an additional layer of grit, and the large chestnuts and acorns can be set individually into place, each one about 5 cm (2 in) beneath the surface. The space allowed for each variety should be no larger than absolutely necessary, to accommodate the seeds sown fairly thickly. Most amateur gardeners would find an area of a 1000 sq cm (155 sq in) ample for their needs; capable of holding something like 50 seedling birches, 20 or 30 beech or ten fine chestnut seedlings. the aim should be to produce a strong stand of fairly closely spaced seedlings, rather than a few plants spread over a large area.

An unexpected problem sometimes causes difficulties. Most of the trees which we grow in our gardens are completely hardy, and we never need give a thought to the possibility that they may be damaged by frosts; many of these produce seedlings which are quite easily killed by freezing temperatures which more mature plants survive unscathed. Seedlings often emerge early in the spring, and need to be protected from frosts in late April and May. Branches cut from conifers and used for covering look authentic and work well but are not

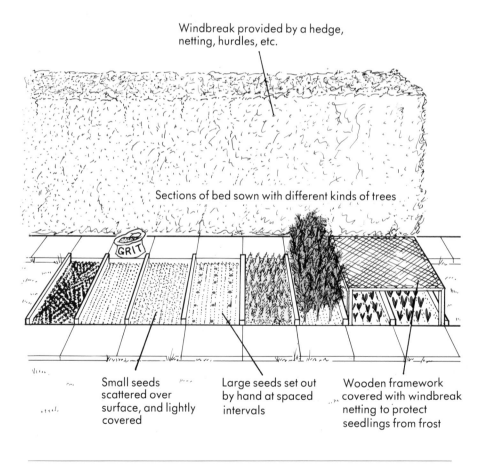

Windbreak provided by a hedge, netting, hurdles, etc.

Sections of bed sown with different kinds of trees

GRIT

Small seeds scattered over surface, and lightly covered

Large seeds set out by hand at spaced intervals

Wooden framework covered with windbreak netting to protect seedlings from frost

Figure 10.1 Many coniferous and broad-leaved trees can be grown from seed sown outdoors in a seedbed. This is a very economical and straightforward way to grow quite large numbers of trees, but they will not develop as rapidly as trees grown under more intensive conditions.

always available, or can be spared without spoiling specimen trees. Sheets of bubble polythene stapled to a light wooden frame are a contemporary, readily available and very easily manipulated alternative, to provide temporary cover when cold nights threaten to do damage.

Seedlings are usually left in their seed beds to grow on through the summer, and can be lifted and lined out the following autumn, unless more rapid results are wanted when young plants lifted as soon as they can be handled, potted up and spoilt a bit will grow extremely rapidly. Conifer seed (p. 66) sown in nursery beds outdoors germinates well, and the great majority of conifers grown for forestry purposes are raised in this way. Their growth during the first year or two is likely to be extremely slow when left in the seed bed, and they do need tending carefully to make sure they are not overwhelmed by weeds. Like many of the broad-leaved trees, they respond dramatically when potted up after the first year and grown under more benign conditions, and when time and space allow, this will help to relieve frustrated gardeners who find it hard to sit back and wait for nature's long, slow cycle to unfold!

The steady growth of many tree seedlings in seed beds usually provokes gardeners to look for ways of hastening their development, rather than hold them back. There are situations where their capacity for slow growth can be an asset, however strange and perverse this may appear to be at first sight, when it is worth while deliberately providing conditions which prevent their growth. Nuts and other large moist seeds cannot be stored effectively for much longer than a year but some, and beech is a famous example, produce crops of seed at irregular intervals, so that up to five or six years may pass from one good crop to another. These intervals cannot be bridged by storing the seeds, as would be possible with other more tolerant kinds, but, by sowing the seeds thickly in a seed bed, and leaving the seedlings to stand closely side by side competing one with another for water and nutrients, they will hold for three or four years, growing slowly, but still able to develop rapidly as soon as they are lifted and lined out individually.

Trees that can be Propagated from their Seeds

Acer	Eucalyptus	Morus
Aesculus	Eucryphia	Nyssa
Amelanchier	Fagus	Paulownia
Arbutus	Fraxinus	Pittosporum
Betula	Fremontodendron	Platanus
Carpinus	Gleditschia	Prunus
Castanea	Ilex	Ptelea
Catalpa	Juglans	Quercus
Cercidiphyllum	Koelreuteria	Rhododendron
Cercis	Laburnum	Robinia
Corylus	Liquidambar	Sophora
Cotoneaster	Liriodendron	Sorbus
Crataegus	Magnolia	Stuartia
Davidia	Malus	Tilia
Embothrium	Mespilus	Ulmus.

Propagation from Cuttings

Many trees can also be raised from cuttings, using methods described previously for shrubs. During recent years commercial nurseries have extended the range produced in this way considerably, and soft tip cuttings, in particular, are now used very successfully, in place of budding or grafting, to propagate some of the Japanese maples, magnolias and some of the varieties and close relatives of the winter-flowering cherry. These are quite specialised techniques during which stock plants are brought into a warm greenhouse during February to procure a very early crop of cuttings, which are potted up as soon as they produce roots and grown on as rapidly as possible through the summer to form the well-developed plants needed to survive their first winter. Although the method depends for success on very good facilities, and quite elaborate care, it may still be of interest to amateur gardeners because the plants concerned are not easily produced by simpler methods, and all are particularly attractive, and expensive, inhabitants of our gardens.

Trees Grown from Tip Cuttings (p. 128)

Acer	*Davidia*	*Magnolia*
Amelanchier	*Hoheria*	*Prunus.*

Trees Grown from Semi-mature Cuttings (p. 130)

Carpinus	*Davidia*	*Paulownia*
Catalpa	*Eucryphia*	*Pittosporum*
Cercidiphyllum	*Hoheria*	*Prunus*
Cotoneaster	*Ilex*	*Rhus*
Crinodendron	*Morus*	*Stuartia.*

Trees Grown from Hardwood Cuttings (p. 133)

Acer	*Fremontodendron*	*Platanus*
Arbutus	*Ilex*	*Populus*
Buxus	*Laburnum*	*Prunus*
Cornus	*Laurus*	*Salix*
Cotoneaster	*Morus*	*Tamarix.*
Eucryphia	*Pittosporum*	

Trees that can be Propagated from Root Cuttings (p. 111)

Ailanthus	*Paulownia*
Catalpa	*Rhus.*

By contrast, a crude and simple method of propagation is used to produce young trees of willows, poplars and mulberries. Each of these can be propagated from cuttings 2 or even 3 m (6½–10 ft) long, prepared by removing entire branches from the parent trees, trimming off the side shoots and inserting them firmly into the open ground, where they are intended to grow. Windbreaks of poplar and willow are very easily produced like this, and grow rapidly into trees, provided the ground within a radius of 1 m (3¼ ft) around each is kept free of weeds for at least two years after they are set out.

Division

Propagation from Layers

Some trees, particularly while they are still immature and growing fast, can be propagated from layers, by making use of side branches which are long enough and trailing enough to be brought into contact with the ground (p. 143). Some kinds of trees are produced in nurseries by layering in large numbers, but this is done in ways which interfere with the development and shape of the tree so drastically that it would seldom be acceptable in a garden. Air-layers, despite their problems, can be used (p. 146) to propagate many trees, particularly when they are growing vigorously and producing long straight shoots, and very often this provides almost the only feasible method of propagation, unless the skills, facilities and equipment needed for budding or grafting are available.

Trees that can be Propagated from Layers

Acer	*Davidia*	*Mespilus*
Arbutus	*Eucryphia*	*Nyssa*
Cercidiphyllum	*Hamamelis*	*Platanus*
Cornus	*Hoheria*	*Prunus*
Corylus	*Magnolia*	*Stuartia.*

Propagation from Suckers

A few trees are extremely easy to propagate from the suckers which they produce naturally. All that is necessary is to dig up suckers as they appear, cut the roots on either side of them, and look after them for a few weeks while they establish themselves. This very simple method will not produce a satisfactory result with trees which have been budded or grafted, in which the roots are part of the stock. Cherries sucker freely, but many are budded on to commercial stocks, which are not worth their place as specimen trees in a garden. Suckers of the golden-leaved false acacia called *Frisia*, hopefully dug up during the winter, will all turn out to have the plain green leaves of the wild green *Robinia pseudoacacia* on whose seedlings this garden form is grafted.

Trees that can be Propagated from Suckers

Nyssa	*Prunus*	*Robinia*
Populus	*Rhus*	*Ulmus.*

CHAPTER ELEVEN

Propagation of Bulbous Plants

Introduction

All plants which produce flowers, and that covers almost everything we grow in our gardens apart from ferns and conifers, have been classified by botanists into two great divisions; the dicotyledons within which are grouped all the broad-leaved plants, and the monocotyledons. The enormous family of the grasses are the epitome of the monocotyledons for those of us who live in temperate parts of the world, and, with their cousins the sedges and rushes, evoke an immediate image for the group. If we dwelt in Central Africa we would be more likely to think of banana plants, and, in many other parts of the Tropics, of palm trees of one kind or another. Grasses, bananas and palms all produce flowers, but none with structures which resemble petals and, though often very beautiful, dramatic even, they lack the brilliance and textural qualities of those produced by more typical flowering plants. However, amongst the monocotyledons there are numerous tribes which produce what appear to be petals, and which yield nothing in beauty, textural quality or brilliance in comparison with any other flowers. They include garden plants of outstanding importance, amongst them daffodils, irises, tulips, arum lilies, the true lilies, orchids and hostas.

A high proportion of these bright monocotyledons produce bulbs, or at any rate, structures like corms and tubers, which look rather like them, all of them are sometimes referred to by gardeners in a broad, all-embracing and uncritical way as bulbs. This takes no account of what is or is not a true bulb, and includes plants like hostas, day lilies, or lilies of the valley, which most of us would not think of as bulbs. There is obviously room for a more precise epithet to describe the group as a whole. Botanists have provided one by coining the gawky and rather inelegant phrase 'petaloid monocotyledon', which may not be attractive but at least indicates clearly that it embraces those monocotyledonous plants in which the flowers have structures which resemble petals. Perhaps, one day, a more endearing way will be found of describing these very beautiful plants. Meanwhile the bulbous plants which follow are those which the botanist would call petaloid monocotyledons.

The group as a whole produces seed abundantly from conspicuous capsules that make the task of harvesting very simple. Like all small dry seeds, they store well (p. 181), and amongst their number are a great many plants whose seeds need no particular skill to encourage them to germinate and produce seedlings.

154

Paradoxically, very few are raised from seed, except by specialists, and most amateurs ignore the possibilities they offer. There is a belief that it takes a very long time for them to produce flowers from seed, and that they are hardly worth sowing on that account. This period can stretch to five or even seven years, but very frequently seedlings will flower 18 months to three years after they are sown, and these are a very practical proposition, particularly as plants can often be produced in large numbers which would be hard to match by other methods of propagation.

Propagation from cuttings is also a possibility even though these bulbs and corms appear to possess nothing that could remotely be called a cutting. Success depends on understanding the differences between the structures of bulbs, corms and tubers and the ways that their component parts correspond to the more familiar, and the more obvious, forms of herbaceous perennials and shrubs. Most bulbous plants contain well-defined groups of cells which are capable of division and regeneration, and these can be used to produce complete new plants from fragments of the parent.

Propagation by division comes naturally to mind when thinking of these plants, but it can be a relatively slow business, when it depends on the natural production of offsets or sub-divisions of bulbs and corms. Even so, it often provides a better return than first impressions suggest. A bulb might, for example, yield only three sub-divisions each year, which would seem quite a trivial rate of increase, but if the rate is maintained year by year for five years, even this slow multiplication builds up to nearly 250 offspring from one original. In most other spheres, particularly financial ones, a yield of 300 per cent *per annum* would be regarded favourably! Some corms, in particular, produce large numbers of cormlets which can be collected and grown on, and many bulbous plants respond to deliberately inflicted injuries by increasing the numbers of offsets they produce.

Seed Germination

Growing Bulbous Plants from Seed

A number of bulbous plants grow naturally in areas where the climate is not particularly severe, or where periods of inclement weather are quite brief. Many of these produce seeds which germinate under simple conditions (p. 56) with no need for special conditioning or other treatments of any kind.

Bulbous Plants whose Seeds Germinate Easily

Agapanthus	*Hosta*	*Nomocharis*
Brodiaea	*Iris*	*Pancratium*
Crocosmia	*Kniphofia*	*Roscoea*
Eremurus	*Libertia*	*Sisyrinchium.*
Galtonia	*Lilium*	
Gladiolus	*Nerine*	

Bulbs and similar structures are amongst the ways that plants protect themselves from harsh conditions, or rest during unfavourable seasons; the production of seeds frequently coincides with the onset of these conditions as the upper parts of the plant wither away, and it retreats into subterranean security. Many bulbs grow in areas which endure extreme heat or drought during the summer, and intense cold during the winter, so that only brief periods of the year are favourable for growth. Under these conditions the bulbs provide a substantial base, which can be built up gradually over the years and filled with storage reserves, from which shoots, leaves and flowers can be rapidly produced when conditions allow.

The seeds produced by bulbs growing in such harsh conditions provide a way for plants to find places where their seedlings could establish themselves, with some prospect of survival, provided they receive their share of the meagre rainfall available. Many species produce seeds, which do not germinate at high temperatures, following the same strategy as annuals from around the Mediterranean (p. 15), but with a slight and vital difference: they also respond to high temperatures as conditioning treatments. The significance of this is that the conditioning treatments are effective only when seeds are hydrated, so that seeds which receive more than average amounts of water are most likely to complete the processes involved during conditioning and to go on to produce seedlings when temperatures drop. In the garden, seeds which need conditioning treatments should be sown as soon as possible after they become ripe, to benefit from the effects of summer warmth. Most will germinate during autumn, although some will delay germination till spring, after they have experienced a second conditioning treatment—this time at low temperatures.

Bulbous Plants that should be Sown as soon as Possible after Harvest

Asphodeline	Hyacinthus	Ornithogalum
Camassia	Ipheion	Puschkinia
Chionodoxa	Iris	Scilla
Crocus	Leucojum	Sisyrinchium
Fritillaria	Lysichiton	Tulipa.
Galanthus	Muscari	
Hyacinthoides	Narcissus	

Many petaloid monocotyledons grow in much less extreme climatic conditions; the lilies of deciduous woodlands provide well-known examples. Others, particularly those which grow as herbaceous perennials, are also woodland plants often found growing naturally under rather moist conditions; and yet others grow amongst grasses and shrubs. The seeds of a high proportion of these produce seedlings exceedingly easily in cultivation provided they are kept warm after they are sown; but at lower temperatures (less than 15°C, 59°F), germination can be very slow indeed or may be delayed indefinitely.

Hostas from Seeds

Hostas, commonly though infrequently called plantain lilies, from woodlands in Japan and a few from China and Korea, behave in this way. Above 20°C

(68°F) they produce seedlings very rapidly, and very few viable seeds remain ungerminated. As temperatures fall to 15°C (59°F) or to 10°C (50°F) or less, the time required for seeds to germinate increases disproportionately so that at the lower temperatures weeks or months may pass before seedlings emerge. These responses to temperature provide a simple but very effective means of regulating germination under natural conditions. Seeds which are shed early, that is during summer, into places which are sunlit and warm, are likely to germinate rapidly and produce established seedlings before the winter. Those which land in cooler more shady places or are shed a little later are not wasted by producing seedlings which are too small to survive, but lie where they are till spring, and germinate as temperatures begin to rise again. Gardeners have no problem therefore; if they want to germinate *Hosta* seeds they make sure that they do so at temperatures above 20°C (68°F).

Two-stage Germination in Lilies

Some of the true lilies germinate rapidly under the same conditions as any bedding plant, and need no conditioning treatment of any kind. Others, particularly those that grow in woodlands where shade prevails and competition can be severe, germinate in two stages, the first providing a foundation for the second. When these lilies are sown at moderate temperatures (15–20°C, 59–68°F) they start by germinating normally. Their radicles emerge and grow into the compost to produce a short root, but, having anchored the seed to the ground, and established contact with nutrients and water, they fail to follow up in the usual way by producing leaves. Instead, a short section between the seed and the root swells to form a minute bulb, building itself up from reserves in the seed and nutrients in the soil. At this point it stops growing, and does nothing more while temperatures around it remain high.

This is a familiar story and very reminiscent of the behaviour of seeds of hellebores, in which an immature embryo develops up to a certain point, and then grows no more until after it has experienced a period of wintry weather. So with these lilies: once winter has passed, and with it a period of several weeks of cold, near frosty weather, and temperatures start to rise in the spring, they make their next move. A leaf emerges from the tiny bulb, followed by more to make a little cluster, which continue to develop through the summer. As successive leaves become larger, more and more reserves are moved back to build up the bulb, and by autumn, though usually not yet large enough to support a flower the following year, it will have made substantial growth, providing a very good foundation from which to compete with the vegetation around it later on.

Figure 11.1   The seeds of some lilies produce seedlings in two stages. At first a tiny subterranean bulb is formed to establish a base, and subsequently small leaves emerge from this bulb and grow up into the light. When the bulbils have been formed the bags containing the partially-germinated seeds should be transferred to a refrigerator for 4–6 weeks, and then opened and their contents emptied into pots half filled with compost, for their leaves to emerge and complete the germination process.

Seeds mixed with moist grit in plastic bags

Seeds of *Lilium martagon*

Bags of seeds and grit put in a warm place while the bulbils develop

First stages in the germination of *Lilium martagon*, showing formation of bulbils

The seeds of lilies which germinate in this way need special treatment before they will produce seedlings. In essence they need:

- A period of about two months at 20°C (68°F) or more; for the radicle to emerge and the bulb to start to develop.
- Six to eight weeks at low temperatures (2–4°C, 36–39°F), to condition the seed, and enable the leaves to emerge later.
- A return to warm conditions (15–20°C, 59–68°F) during which the seeds complete their germination, and grow on to form small plants which will build up the bulb.

These sequences are usually easier to provide under entirely artificial conditions than by relying on natural cycles produced by the changing seasons. Seeds can be mixed with moist grit or vermiculite, and packed into strong polythene bags before being put somewhere where they can be kept warm for up to two months at temperatures close to 20°C (68°F). Every now and then they should be looked at to find out if they have started to germinate by producing roots; the emergent radicles may be small, but they are quite conspicuous, and not hard to spot if they are present. Subsequently, bead-like swellings will appear on the roots which are the first stages in the development of bulbs, and when these stop growing the time has come to move the bags containing the seeds to different conditions. The low temperatures needed to condition the bulblets, and enable the leaves to emerge later, can be provided very easily, by putting the bags in a refrigerator for about six weeks. After this they should be moved back into the warm and as soon as green leaves start to sprout from the tiny bulbs they can be pricked out into small pots containing potting compost, and grown on in a warm greenhouse or a frame.

Lilies that Germinate under Normal Conditions

Lilium amabile	L. formosanum	L. regale
L. concolor	L. longiflorum	L. sargentiae
L. davidii	L. pumilum	L. tigrinum.

Lilies that Germinate in Two Stages

Lilium auratum	L. japonicum	L. rubellum
L. canadense	L. martagon	L. speciosum
L. columbianum	L. monadelphum	L. szovitsianum.
L. hansonii	L. pardalinum	

Conditioning Treatments

Some bulbous plants produce seeds which are reluctant to germinate and may not do so for months or years, if at all, after they are sown. Some members of the family Liliaceae, many of the wild onions, and alstroemerias or Peruvian lilies, behave like this. Some will not germinate at all until their seed coats are damaged in some way; others germinate much better after this has happened.

Usually the effects produced by chipping or tearing the seed coats must be co-ordinated with other physiological changes resulting from conditioning treatments.

The need for some kind of damage to the seed coat recalls the behaviour of many seeds produced by members of the pea family (p. 50), and other dicotyledonous species which produce seeds with hard, impermeable seed coats. All make sure that the embryos within the seeds cannot take up the water they need in order to germinate, until the seed coat is fractured. The seeds produced by Peruvian lilies, wild onions, some irises, and other plants which share similar germination responses, are leathery rather than hard, and are able to take up water through their intact seed coats without difficulty. The effects of damaging them probably has nothing to do with the way the seeds take up water, but may allow oxygen to diffuse more rapidly to the embryos, by removing the effects of membranes which previously restricted the supply available, especially since the ruptures in the seed coat usually work most effectively when they are made very close to the positions where the embryos lie.

Bulbous Plants that may Require Complex Conditioning Treatments and/or Damage to their Seed Coats

Allium	Curtonus	Nomocharis
Alstroemeria	Erythronium	Ophiopogon
Anthericum	Hemerocallis	Polygonatum
Cardiocrinum	Iris	Ruscus
Colchicum	Lilium	Tulipa.

Propagating *Alstroemerias* from Seed

Peruvian lilies grow in parts of the southern half of the Andes in places where late summer may be a period of drought, but where winters are not so severe as to make survival of seedlings impossible. The most favourable season for the emergence of seedlings is the autumn. Seeds fall from their large capsules during the summer often fairly early on, and lie in the ground during the remainder of the summer when soil temperatures are high. During this time the seeds may be invaded by fungi or bacteria, or attacked by insects, small animals and birds, and some will suffer damage to their seed coats in the process. These seeds are most likely to germinate when temperatures fall in the autumn, leaving most of the undamaged seeds to lie on in the soil for another year or longer.

Seeds of these plants can be raised by putting them, mixed with moist vermiculite, into a polythene bag, and keeping them in an electrically heated propagator or some other place where a temperature of about 25°C (77°F) can be maintained for six to eight weeks. The seeds must then be temporarily separated from the vermiculite; they are quite large seeds and this is not difficult. The slightly tricky bit comes next. A small part of the seed coat and the tissue beneath it has to be removed from each seed from a point immediately above the place where the embryo lies. Fortunately, in *Alstroemerias*, the place to make the attack is very conveniently identified by a shadowy, dark spot on

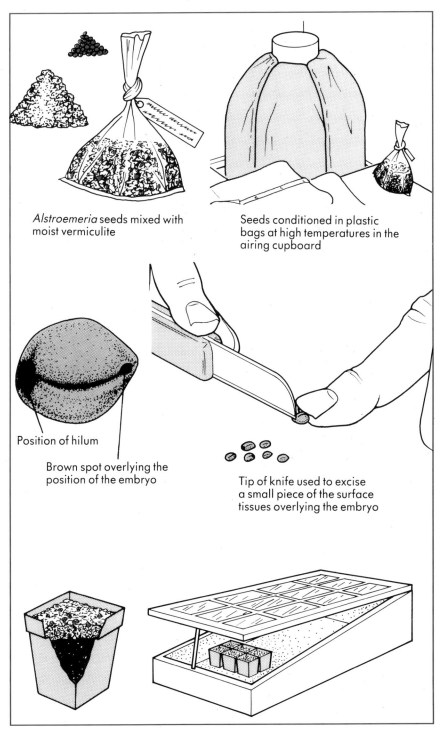

Alstroemeria seeds mixed with moist vermiculite

Seeds conditioned in plastic bags at high temperatures in the airing cupboard

Position of hilum

Brown spot overlying the position of the embryo

Tip of knife used to excise a small piece of the surface tissues overlying the embryo the seed coat, which can be excised with the pointed tip of a knife. The treated seeds can then be returned to their vermiculite, spread over the surface of a pot containing compost, and put in a cool place where they will produce seedlings. They will germinate most rapidly at temperatures around 10°C (50°F); much above this and seedlings will not emerge, much below it, and although they will germinate, they do so rather slowly. The best plan is to organise the time when a start is made so that the treated seeds are ready to go out into an unheated greenhouse or a cold frame during late February or early March, which provides them with the low temperatures they need to germinate, and enables the seedlings to develop without artificial heat under almost natural conditions until they start to die down naturally after mid-summer. At other times of the year 10°C (50°F) may not be a very available temperature, and an alternative is to put the treated, and sown, seeds into a refrigerator for a week or ten days then take them out and grow them on under warmer conditions. There will be no visible signs of germination after so short a time in the refrigerator but the seeds will have accomplished the essential first steps towards germination and complete them after their transfer.

Propagation from Cuttings

Comparisons of Bulbous and Other Plants

True bulbs, but not corms, rhizomes or tubers which may resemble them superficially, but are constructed differently, are very much like the buds of more typical plants. They are formed from a number of enfolding layers, derived from leaf bases, which are attached to a compact disc of tissue known as the basal plate. In other more typical plants the stems elongate as the plants grow so that the leaves on them are spaced more or less widely apart. The stems of bulbs do not elongate in this way but spread sideways to form the basal plate to which the leaves, separated by almost imperceptible spaces, are joined. This arrangement becomes very significant when gardeners wish to propagate bulbs: as in almost all other plants, cells capable of division are found at places where leaves and stems join. These small centres of dividing cells are the origin of the offsets which bulbs produce naturally, usually rather slowly. Potentially they are capable of producing far more offsets than they do, and this natural ability to reproduce can be stimulated by cutting bulbs up, or pulling them apart, in such a way that each fragment contains some of the regenerative tissues.

Any action which wounds a bulb, corm or rhizome, such as slicing with a knife, pulling leaf scales away from the basal plate or cutting rhizomes into

Figure 11.2   The seeds of almost all kinds of Alstroemerias *require special treatment before they will produce seedlings. After a conditioning treatment at high temperatures the seed coat must be lacerated close to the place where the embryo lies. After this the seeds will germinate rapidly at temperatures below 10°C. A greenhouse bench during the winter or a ventilated cold frame in spring provide the right conditions. During the summer a few days in the refrigerator will get the seeds going.*

sections, can stimulate cells in the regenerative tissues to divide and, at the least, repair the damage done: it is also likely to result in the production of small bulbs, corms or offsets. Similar attempts to cut up, or injure, tubers will not result in the production of new plants.

True Bulb-producing Plants

Allium	*Galtonia*	*Nerine*
Brodiaea	*Hyacinthoides*	*Nomocharis*
Camassia	*Hyacinthus*	*Ornithogalum*
Cardiocrinum	*Ipheion*	*Pancratium*
Chionodoxa	*Iris*	*Puschkinia*
Crinum	*Ixiolirion*	*Scilla*
Erythronium	*Leucojum*	*Sternbergia*
Eucomis	*Lilium*	*Tulipa.*
Fritillaria	*Muscari*	
Galanthus	*Narcissus*	

Corms and rhizomes sometimes resemble bulbs, but are formed in quite a different way. Like bulbs, they are made up of tissues derived from stems and leaves, but, unlike them, almost their entire bulk consists of stem tissues, and the leaves exist as mere remnants, reduced to dry scales. Corms, produced for example by crocuses, are usually covered by these dry scales, which form a protective coating several layers deep over their entire surface. These remnants of leaves are less conspicuous in rhizomes, and are often reduced to almost insignificant papery filaments marking the positions of the nodes. Whether conspicuous or barely discernible these structures still count as leaves, and the places where they meet stems can be relied on to provide regenerative tissues capable of division to produce new plants.

Corm-producing Plants

Brodiaea	*Crocosmia*	*Curtonus*
Colchicum	*Crocus*	*Gladiolus.*

Rhizome-producing Plants

Acorus*	*Houttuynia*	*Ruscus*
Asphodeline	*Iris*	*Smilacina*
Calla*	*Polygonatum*	*Trillium*
Convallaria	*Rhodohypoxis*	*Zantedeschia**.

* Plants that grow as aquatics.

Tubers are yet another form assumed by plants which can look very like bulbs or rhizomes. Their similarity to the latter is so close that, very often, no distinction is made between them. In fact, tubers are swollen roots produced by distinction is made between them. In fact, tubers are swollen roots produced by plants to store food reserves; rhizomes are swollen stems, which may store food reserves, but also play an important part in the ways plants reproduce and survive naturally. Distinguishing between them becomes vitally important in propagation. On the surface of a tuber no buds exist which could be used to produce new plants, although a few may often be found clustered around stems which emerge from the top of the tuber itself. Rhizomes have buds distributed over their surfaces wherever a node occurs, and on a potato 'tuber' these show up as the eyes, and these have already been described (p. 98) as a potential source of plants.

Tuber-producing Plants

Alstroemeria	Arum	Sagittaria*.
Arisaema	Hemerocallis	

* Aquatic.

Quite a large number of these bulbous plants, less ambiguously referred to as petaloid monocotyledons, produce no bulbous structures at all. They grow like herbaceous perennials by producing crowns which multiply from year to year to form expanding clumps. A few, including some species of *Hemerocallis*, may also produce tubers, and some like the red hot pokers (*Kniphofia*) are often more or less evergreen. Almost all are very easy to propagate by division during July and August when they produce very strong anchor roots from the base of their crowns.

Bulbous Plants that are Similar to Herbaceous Perennials

Agapanthus	Hosta	Pontederia*
Alisma*	Kniphofia	Roscoea
Anthericum	Libertia	Sisyrinchium
Butomus*	Liriope	Tradescantia
Eremurus	Lysichiton*	
Hemerocallis	Ophiopogon	

* Plants that grow as aquatics.

Propagation of Lilies from their Scales

The scales which compose a lily bulb are packed loosely together, and are easy to remove without cutting them apart. Each can be pulled slightly sideways and downwards so that it breaks off at the point where it joins the basal plate. Provided they are then kept alive and continue to function, cells along their broken edges will divide to repair the damage done. These repairs are not limited to renewing damaged tissues; some cells will continue to divide and go on to produce bulbils along the broken surfaces—usually one or two, but occasionally more.

Scales can be removed from lily bulbs at any time of the year, but the most convenient times are when they are dug up to be replanted, and when new

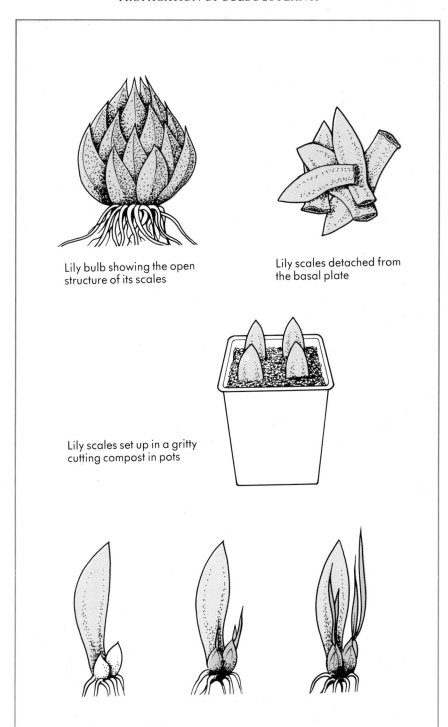

Lily bulb showing the open
structure of its scales

Lily scales detached from
the basal plate

Lily scales set up in a gritty
cutting compost in pots

batches of lilies have been bought. The very old, semi-rotten and shrivelled remnants of scales on the outside are thrown away, and a proportion of the remaining mature scales removed and used to produce bulbils. At least a third can be taken off without ruining the bulb's prospects of producing flowers. It is important to break them off gently and carefully so that each separates at the junction with the basal plate; these are in fact leaf cuttings (p. 95), and the inclusion of regenerative tissue in each is just as essential as it is with other plants propagated in the same way, like ramondas.

As soon as they have been removed the scales should be put into a polythene bag with a small amount of fungicidal dust (Benomyl for example) and lightly shaken to protect them from moulds. Each one is then stuck into a gritty compost, so that about a bare third protrudes above the surface. They can be left to develop, without artificial warmth, in an unheated greenhouse or a cold frame, but will progress much more quickly and with greater prospects of success if they can be kept at 10–15°C (50–59°F). Small bulbs form along the cut surfaces, and produce grass-like leaves which emerge above the compost. As soon as these leaves begin to appear they should be drenched with a liquid feed, and further feeds given at monthly intervals encourage the young plants to develop. Usually it is best not to rush to pot the young plants up individually. If this is done while they are still very small, they need careful and skilful attention to keep them going, and it is much easier to leave them in their well-drained cutting compost and rely on regular feeds to keep them going until they are established, and their bulbs are well-developed. Their leaves die down as winter approaches, and the bulbs should be left undisturbed in a cold frame protected from the weather, and from mice and slugs with a winter living to make, until the first signs of young leaves appear the following spring. This is the time to separate the young plants gently and pot each up separately, so that they can grow away and develop strongly through the summer.

Twin-scaling Bulbs of Daffodils and Narcissi

Narcissus bulbs are densely compact, and each of the leaf bases, which form the scales, overlies and enfolds another to form a closely knit mass, which cannot be prised apart like a lily bulb. They reproduce naturally by offsets which grow from the regenerative tissue around the basal plate and appear as side shoots, which form new bulbs: their rate of production varies from one variety to another, but is always steady rather than rapid.

A daffodil bulb growing in the garden and cut in two by a spade does not usually die. The halves, like the proverbial worm, but much more successfully, repair themselves and soon form two complete bulbs. If divided more deliberately into four, or six, or eight segments each of these can also reconstruct a

---

Figure 11.3   True lilies produce bulbs with large, loosely attached scales. These can be snapped off at the point where they are attached to the basal plate, and will then produce bulbils along the broken edge at the bottom of the scale. The pots containing lily scales can be set out on a heated bench, in an unheated greenhouse or in a coldframe. Sooner or later, depending on the temperature, bulbils will form along the broken edges followed by grass-like leaves.

complete bulb, provided that its remnants include sections of the regenerative tissue at the junction of the basal plate and the leaf scales. The extent to which a bulb can be sub-divided and used to produce more bulbs in this way is only limited by the skill, needs or greed of the propagator. Repeated sub-divisions can be made and, with practice, it is technically possible to produce several hundred fragments from one bulb, each consisting of no more than a shaving from two leaf scales united by a tiny piece of the basal plate. These slivers are known as twin scales and they are used to build up stocks under commercial conditions, when a very few bulbs are available but enormous numbers are needed.

Amateur gardeners usually have more modest needs, and these can be satisfied by relatively coarse divisions of the bulbs to produce eight, 16 or, at most, 32 sections from each. The best time to cut the bulbs up is during July and August after their tops have died down, and when very little active growth is in progress. They are laid on their sides and sliced from top to bottom with a sharp knife to produce sections, which should be immediately put into a polythene bag with a little fungicidal dust, and shaken about to give each a protective coating. They are then mixed with between five and ten times their volume of moist vermiculite and packed into polythene bags.

Daffodils are cool creatures, and this is one occasion when high temperatures do not speed things up. The ideal temperature for the development of the young bulbs is about 10°C (50°F), and the rate at which they progress will be reduced if temperatures rise much above 15°C (59°F). A cellar provides the right conditions, alternatively a well-insulated and shaded shed will suffice, or a corner of the garden in the shade of dense evergreen shrubs. If enough bulbs are being propagated to warrant the effort, a practical solution is to dig a small pit in a shaded corner of the garden and store the bags in it under a well-insulated covering made from sheets of expanded polystyrene.

From time to time the bags should be opened and examined to look for leaves emerging from the sections. When these appear the fragments of bulbs are set out in rows in small containers in a very well-drained, gritty compost and grown on in a cold frame through the late winter and spring, and until the leaves die down in due course during the following summer.

Provided that greed was tempered by discretion when the bulbs were cut up, so that the sections made were not extremely small, little bulbs should be present at the end of the first season, which can be knocked out of their containers, and set out in a nursery bed. Some will produce flowers after another season's growth, and the remainder after a further year. Before cutting up any bulbs it is worth reflecting that if very large numbers of tiny sub-divisions are made each will take a long time to grow and produce a flower, and will depend on skilled attention to keep it alive. It may well be that 16 quite substantial sections, which come into bloom in less than three years, are a better bet than 100 or 200 which may grow for four or five years before they flower, if they survive to do so.

Propagating Bulbs of Hyacinths and Fritillaries

Hyacinths and fritillaries both produce bulbs composed of a few heavy, succulent leaf scales. When left to themselves they produce offsets quite slowly, but, if injured, can be stimulated to multiply more rapidly. Hyacinths have a well-developed and very conspicuous basal plate, and, when this is damaged in any way, small bulbs tend to be produced close to the place where it was injured. The easiest way to stimulate bulbil formation is to make cuts across the base of the bulb during July and August with a sharp knife. Another method is to gouge out most of the basal plate to expose the tissues at the base of the leaf scales. The instrument usually recommended is a sharpened teaspoon, which sounds most effective, but could have savage consequences if used later for more usual domestic purposes. A grapefruit knife does the trick. After mutilation the bulbs are set out individually, upside down in a warm dry place, while they heal, and as they do so small bulbils develop on the cut edges, and around the bases of the leaf scales.

These bulbils are picked off and grown on in a gritty compost in small containers preferably within the shelter of a cool greenhouse. The bulbs take three or four years to grow large enough to produce flowers, and the method is often more or less dismissed as of little interest to amateur gardeners. However, it is an interesting and effective method of propagation, and one which could be worth a try if the bulbs used are some which have produced flowers in pots for the house. These tend to do rather badly when planted out in the garden, and often more can be gained by using them as stock plants from which to produce new bulbs.

The resting bulbs of fritillaries, particularly the large, odoriferous bulbs of the crown imperial, can be sliced vertically, when they have died down after flowering, into segments, which are set up like lily scales in a gritty compost, but with each one buried with its top just beneath the surface. They should be treated with a fungicidal dust and kept barely moist, but not dust dry, while they repair the injury done to them, and start the development of new plants.

Propagation of Corms and Rhizomes

The spherical corm produced by a crocus or gladiolus may not look much like one, but it is in all essentials just a dumpy little shoot. The roots are at the bottom, there is an apical bud at the top, which normally develops to produce flowers and leaves, and in between a section of stem bears leaves—even though these are brown and scale-like and cannot function like them. Rhizomes have a very similar structure, and it takes less imagination to think of them as shoots. In both cases the essential point is that development of side shoots, which in this case means the production of offsets, is controlled by the apical bud, which holds back the outgrowth of buds in the axils of the leaves below it.

When the apical bud on a normal shoot is destroyed or removed during pruning it can no longer control the growth of the buds beneath it, and these grow out as side shoots and eventually take its place. In the same way, the dominant buds at the top of crocus or gladiolus corms, or at the end of trillium rhizomes, prevent almost all the other buds from developing into offsets. When these dominant buds are cut out with the point of a knife, buds lower down which may be mere pinpoints will respond by renewed activity, and eventually develop into small shoots. These form offsets which can later be removed, and set out individually to grow into flowering plants. The effect is seen quite often in crocuses growing in the garden after mice have eaten the terminal bud, and a mass of grass-like leaves develop from buds lower down the corm. Removing what seems to be the very heart of the plant can call for considerable courage when precious things like trilliums are subjected to the knife, especially when the subsidiary buds on which everything depends are not very much in evidence. In fact it is an unusually safe form of propagation, in which the chances of a successful result are very high indeed.

Division

Increasing Tulip Bulbs

Tulip bulbs sub-divide naturally after flowering, into smaller bulbs, some of which will be large enough to flower during the following year; others, too small, will increase in size for a year or two before doing so. Most varieties and species tend to decline in gardens, and, unless they are skilfully managed, end up after a few years producing only a few small flowers and numerous small, non-flowering bulbs. The depth of planting influences the behaviour of the bulbs, and premature decline can be caused by planting too close to the surface. Shallow planting, and annual lifting and sub-division, encourage rapid increase in the number of bulbs, but few will grow in size to become large enough to flower well. Large, flowering-sized bulbs can be encouraged to develop by deep planting (at least 20 cm, 8 in) and the removal of flowering stems from undersized bulbs as soon as they become visible. Some of the species tulips grown in our gardens produce 'droppers'; bulbs which are formed well below the level of the other bulbs, perhaps 30 cm (1 ft) below the surface. As the remaining bulbs gradually decline and cease to flower or disappear, these deep set bulbs increase in size and keep the colony going.

Dividing Narcissus Bulbs

Daffodil bulbs which are left in place gradually build up into large congested clumps, containing a few bulbs which continue to flower and large numbers of undersized, non-flowering bulbs. Provided they are healthy the absence of flowers is simply due to competition within the clump, so that many individuals fail to obtain enough nourishment to sustain their development. This is easily remedied. Clumps should be lifted as the leaves die down, and the bulbs separated, and graded into those large enough to produce flowers, and the rest. In September the large bulbs can be planted in the garden and the smaller ones lined out in rows to grow on for a year or two.

Usually the smaller bulbs increase in size rapidly once they are planted separately and produce flowering-sized bulbs within a year or two. Batches of bulbs which remain blind for two years or longer, or which produce flowers very sparingly, should be looked on with suspicion. The most probable cause is infection with one or more virus diseases. It is possible to eliminate these by keeping the bulbs for long periods at high temperatures, but these treatments are not practicable unless the stock involved is unusually valuable for some reason, and bulbs of daffodils which persistently fail to flower should usually be dug up and burnt.

Saving Gladiolus Cormels

Almost all gladioli produce little corms prolifically from the places where old disintegrating leaf scales are joined to the base of the flowering corms. These cormels can be detached when the plants are lifted in the autumn, and stored, hung up in paper bags, in a dry, airy and frost-proof place. They are looked upon as great delicacies by mice, and the way they are stored must take this into account. The following spring they can be set out in a grid pattern, at 3 cm (1¼ in) centres, in large pots, or lined out in a nursery bed, sowing them like peas in shallow drills, and grown on through the summer. The following autumn they can be lifted, cleaned and stored until spring.

Most will flower after growing for two years, but any which flower precociously should have their flower stems removed to encourage the corms to develop. All gladioli tend to produce seeds abundantly, and as these compete for resources with the developing corms the flowering stems should usually be removed as soon as the flowers fade, unless the seed is to be collected and used. Seedlings will not come true, but they do grow to flowering size quite quickly, and produce an astonishing variety of flower forms. They are easy and interesting to grow and worth the effort from time to time.

Propagation of Lilies by Division

Almost all lilies can be produced by simple division of their bulbs which split up, after flowering, to produce two or more bulbs where one was before. Some like the tiger lily are very productive and may yield as many as five new bulbs from a single parent.

Many different kinds of bulbs can be propagated by cutting them or injuring them so that offsets or bulbils are produced at the junctions of the modified leaves and stems of which a bulb is composed. Some lilies take this a stage further, and produce bulbils spontaneously along their flowering stems in the angles of their leaves. Usually these develop after the flowers have faded, and can be collected just before they would drop off naturally, as the leaves change colour and fall in the autumn. These bulbils can be stored through the winter in polythene bags of almost dry peat, or set out as soon as they are gathered in pots or deep containers, overwintered in a cold frame and grown on through the following summer. Most will grow to produce a bulb large enough to flower during their third summer.

Some lilies which do not produce bulbils naturally can be persuaded to do so by wrenching the flowering stems away from the bulbs as soon as the flowers fade. Any developing seed capsules are then cut off, and the stems are buried at an acute angle in shallow trenches of gritty compost so that only about one-half protrudes above ground level. They will remain alive and their exposed leaves will continue to function, and, since they possess neither seed capsules nor a parent bulb to which resources would normally be transferred, they can often be persuaded to produce small bulbils in their place.

Lilies that Produce Bulbils Spontaneously

Lilium bulbiferum	*L. sargentiae*	and a great many garden hybrids.
L. candidum*	*L. testaceum**	
L. myriophyllum	*L. tigrinum*	

* Induced by wrenching and burying their stems.

Lilies that can be Propagated from Underground Bulblets

Lilium auratum	*L. regale*	and many garden hybrids.
L. henryi	*L. speciosum*	
L. longiflorum	*L. tigrinum*	

171

CHAPTER TWELVE

Seed Collection and Storage

Introduction

We mostly think of seeds as being small, hard, dry and a shade of brown. They are unchanging and appear to be inert. They are formed, sometimes by the dozen, sometimes by the million, in the fruits which plants produce. Fruits in colloquial usage are juicy, succulent and sweet: for the botanist, the word fruit describes any of numerous kinds of dry or succulent structures which develop from the carpels of a flower. The majority of botanical fruits are not succulent at all, but dry, and a surprising number of the succulent objects which everyone thinks of as typical fruits are dismissed by botanists as false fruits, such as strawberries, pineapples, plums and pears. Perversely, so it seems, others we would think obviously are seeds—like the 'seeds' of lettuces, celery and French marigolds—are referred to by botanists as fruits.

Gardeners need not be concerned with the definitive niceties which distinguish seeds; from their point of view anything which appears to be a seed functions as a seed, and can be treated appropriately, and in this chapter the word is used indiscriminately and from a gardening viewpoint. However, seeds are divided physiologically into two major groups: those which are dry and for the most part small; and moist seeds, which tend to be large, and sometimes, like the double coconut, very large indeed. The two kinds are usually easy to tell apart, and they behave in such different ways that it is worth understanding the differences between them.

Moist seeds form a minority group. They include most of the things we call nuts, and other obviously analogous structures like acorns. Most are large, and filled with fleshy tissues packed with storage reserves. When they become dry their fleshy tissues lose water; they shrivel and then they die. They are not long-lived under any conditions.

Dry seeds are quite different, and make up the great majority of the seeds which gardeners are accustomed to handling. They are nearly always small—peas and beans are some of the largest—and they can tolerate becoming extremely dry without suffering harm. Although they appear to be inert, they are, nevertheless, responsive to their surroundings and can deteriorate quickly under unfavorable conditions. On the other hand, under the right conditions, dry seeds can survive for a very long time; decades readily, centuries and even thousands of years occasionally. The essential difference between dry and moist

172

seeds is that the former can survive desiccation without damage, whereas the latter die quickly as they wrinkle.

Seeds that fail to germinate when they are sown are most likely to have failed for one good reason: they are dead. This may appear to be so obvious that to state it is merely superfluous, but there are no discernible differences in the appearance of live and dead seeds, and gardeners are repeatedly tempted to take chances with seeds they know have been around a long time, or about which they know nothing at all. Gardening lore teaches us that quite a number of seeds should be sown fresh from their seed capsules, not kept until the following spring. Primulas and the blue poppies are amongst their number, but both produce more seedlings, more easily, if their seeds are stored well and thus remain alive.

Taking a chance with a doubtful packet of seed scarcely matters when annuals or vegetables are being sown, because these come up very quickly, and they can easily be replaced as soon as the failure becomes obvious. When alpines, shrubs or some hardy perennials are being sown, which need lengthy conditioning treatments before they will germinate at all, it can be very frustrating to discover, eventually, that time has been wasted by using dead seeds. By the time all hope of seeing any seedlings has been finally abandoned it will almost certainly be too late to retrieve the situation, and very often the seed is of a kind which is not too easy to get hold of and its loss cannot be easily made good. Seeds of this kind are worth the extra trouble required to make certain that they remain in good condition until the time comes to sow them. Many of the most attractive, or unusual, plants we would like to grow produce seed rather sparingly, or intermittently, so that progeny of one harvest may have to be used to provide seeds for sowing over several years. There is no difficulty at all in keeping these seeds alive provided a few simple precautions are taken when they are collected, and they are cared for in the right way afterwards.

Collecting and Harvesting Seeds

Harvesting Seeds from Dry Fruits of Perennial Plants

Many plants produce enormous numbers of seeds. They provide an ideal way to introduce new plants to gardens either from the wild (but only when this will not reduce the population) or from plants in our own or other gardens, from which we have been given permission to pillage. They are the simplest possible way to transport plants, being small and relatively tolerant of casual treatment.

Apart from a very few exceptions, all seeds should be collected as they become fully mature, and failure to do so makes such a difference to the vigour and storage life of collections of seed that this should be made a rule that is seldom broken. The correct time is nearly always easy to judge from the appearance of the fruits in which the seeds develop. As they mature they change colour from green to yellow, and then, if they are dry fruits, become obviously dehydrated, sere, and brown or straw-coloured. Fleshy fruits change in a different style and usually become brightly coloured and succulent. There are few problems involved in collecting seeds of most dry fruits, though sometimes

173

persistence and vigilance is needed to gather the seeds before the plant disperses them to the four winds. A great many seeds are contained in capsules, one of the most abundant of all kinds of dry fruit, and one of the easiest of all to collect from successfully. The seeds of poppies, pinks, meconopsis and campanulas are held securely in deep cup- or vase-shaped vessels and can be gathered in large quantities by cutting off their capsules as they start to open, and holding them upside down in paper bags, envelopes or plastic flower pots lined with newspaper.

Other plants may need more careful selection; the follicles along the spike of a delphinium, the ranks of capsules produced by foxgloves or the successive whorls of capsules on a candelabra primula, are all easy to find and to harvest, but these ripen successively over a period of several weeks. The entire stems with their mixture of ripe and unripe fruits and seeds should not be cut off, but each capsule or follicle should be removed individually as it matures. Some species need great care and repeated visits, during each of which only a very few seeds may be obtained. These include violas, the members of the pea family, cranesbills, and the large, aggressively spiny spikes of the acanthus. These produce fruits which are the artillery of the plant world and each has developed ballistic methods of throwing their seeds across the countryside as they ripen. All should be collected by removing individual fruits soon after they start to change colour, a day or two before they finally mature and sling their contents away. The ripening fruits can then be enclosed securely in an envelope or some other ventilated container, from which the seeds will not be thrown out as the fruits complete the processes of ripening.

Collecting Seeds from Trees and Shrubs that Produce Dry Fruits

A great many trees and shrubs produce seeds in dry fruits, and capsules or other structures are frequently found which hold large numbers of small seeds and can be collected without difficulty. These should be removed and prepared for storage in exactly the same way as the small dry seeds of annual and perennial herbaceous plants.

Shrubs that Produce Dry Fruits with Small Seeds

Abutilon	*Cytisus*	*Helichrysum*
Callistemon	*Deutzia*	*Hydrangea*
Carpentaria	*Disanthus*	*Hypericum*
Caryopteris	*Eccremocarpus*	*Indigofera*
Celastrus	*Enkianthus*	*Kalmia*
Chimonanthus	*Eucryphia*	*Lavandula*
Cistus	*Euonymus*	*Lavatera*
Clematis	*Exochorda*	*Phlomis*
Clethra	*Fremontodendron*	*Pieris*
Colutea	*Gaultheria*	*Piptanthus*
Coronilla	*Genista*	*Pittosporum*
Cotinus	*Halesia*	*Rhododendron*

| *Rhus* | *Schizophragma* | *Ulex.* |
| *Salvia* | *Spartium* | |

Trees that Produce Dry Fruits with Small Seeds

Acer	*Eucalyptus*	*Platanus*
Ailanthus	*Eucryphia*	*Populus*
Alnus	*Fraxinus*	*Ptelea*
Betula	*Gleditschia*	*Rhododendron*
Buxus	*Hoheria*	*Rhus*
Carpinus	*Koelreuteria*	*Robinia*
Catalpa	*Laburnum*	*Salix*
Cercidiphyllum	*Liquidambar*	*Sophora*
Cercis	*Liriodendron*	*Stuartia*
Crinodendron	*Nyssa*	*Tilia*
Embothrium	*Paulownia*	*Ulmus.*

Seeds should always be collected in fine weather, and when the plants are as dry as possible. This may be a counsel of perfection in some seasons, and is the sort of smug advice that causes fury when the rain is pouring down, and time is running short, but it is sufficiently important to justify a real effort. Apart from being a much more pleasant task—and collecting seeds from soggy plants can be one of gardening's more dispiriting jobs—seeds collected when wet are not easy to dry out properly, and are very likely to be attacked by moulds, especially when numerous seeds are packed closely together. The second major principle is to remember that it is seeds which are wanted, not seeds plus bits and pieces of assorted plant material, and odds and ends of soil, sand or dead leaves. Seed collecting is not usually difficult, but one of its skills lies in finding simple ways to keep every collection made as clean as possible. Collections of seeds, mixed up with miscellaneous fragments, can be gone through and cleaned up later, but this is time-consuming and very tedious. If they are not removed, and in due course are sown with the seeds, these dead fragments are likely to be attacked by grey mould or other fungi, and can act as sources of infection which hazard the survival of the seedlings around them.

From the moment they are collected all these dry seeds must be kept in containers, open to the air, somewhere dry and well-ventilated, while they complete their ripening processes and lose any surplus moisture. Paper envelopes, with their corners sealed with masking tape, plastic pots lined with newspaper or polystyrene cups all make suitable containers. Polythene bags are to be avoided, as are stoppered bottles, plastic boxes, or tins. All produce air-tight seals which trap water from the newly harvested seeds, and produce the humid conditions in which moulds flourish.

Figure 12.1   Many plants produce small, dry seeds, which can easily be collected and used to grow more plants for the garden. The most important points to remember are to make sure that seeds are thoroughly dry before they are put into packets of any kind, and to avoid mixing the seeds with other bits of plant, soil or sand while collecting them.

Spikes and capsules of *Linaria, Antirrhinum, Dianthus barbatus* and *Lilium martagon* ripen in sequence from top to bottom

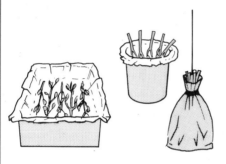 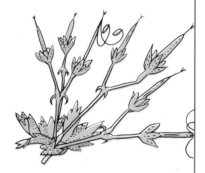

Spikes of *Antirrhinum* being dried in trays and plastic pots lined with newspaper, and in paper bags. Freshly harvested seeds must never be enclosed in watertight containers, in which high levels of humidity build up

Plants like Geraniums, which produce ballistically-propelled seeds should be enclosed in paper bags while drying

Seeds of Hollyhocks being cleaned by hand to remove plant rubbish and bits of soil before packeting or sowing

If seeds can only be collected when they are soaking wet, they must be dried as quickly, but as gently, as possible, bearing in mind that, in any kind of enclosed space, the combination of heat and water produces the hot and steamy conditions that seeds are least able to endure. Very wet seeds should be spread out in a thin layer on a sheet of kitchen paper, and dried; the airing cupboard, a shelf above a radiator, or a spot close to the central heating boiler are ideal places. The heat applied should be gentle and persistent for a few days, and not the hot blast of an oven, or the direct effects of sunlight in a greenhouse. Seeds in brown manilla paper bags hung up in a greenhouse will be shaded from the sun's rays, and will dry out very satisfactorily.

A week or so after collection, the seeds can be separated from the capsules, and any other parts of the parent plant gathered with them. This should be done as gently as possible. Usually seeds can be shaken or vibrated free, and it should not be necessary to break up the capsules containing them. It may be tempting to do a really thorough job by breaking up the fruits containing the seeds to make sure that not a single one is missed. This is very often misapplied effort: the majority of the seeds which do not separate easily are immature or partially developed, and are unlikely to be capable of germinating. Their presence reduces the value of a collection and may do actual harm by providing soft targets for fungal infections when the seeds are sown. Some plants like flaxes (*Linum*), medicks (*Medicago*), eucalyptus (*Eucalyptus*) and the bottle brushes (*Callistemon*) do produce fruits in which the seeds are retained naturally, and these must be broken up, or treated in some way to release their seeds.

Samples of seed which are contaminated with dust, broken bits of plants, immature seeds and other rubbish should be cleaned before being packeted and stored, otherwise it is impossible to tell how much seed there is, or even whether there are any seeds at all. A nylon kitchen sieve can be a great help to get rid of dust and hairs, and it is often possible to make progress by winnowing the seed—put everything in a small cardboard box, and puff away the lighter particles; one soon learns how to blow the lighter particles away from the dense, fully developed seeds. This is an exceedingly effective method but it brings the operator into close contact with dust and hairs among the seeds, so hay-fever sufferers are advised against using this method. By far the most effective way to clean small samples is by hand picking: pour the seeds on to a sheet of white paper and laboriously separate the good from the bad with the point of a small knife. This is so tedious that all who intend to collect seeds should be introduced to it at an early stage to impress upon them the importance of avoiding needless contamination whenever possible!

Clean samples of seed should be put into brown manilla envelopes, sealed at the corners with masking tape, and clearly identified, by writing on them the names of the plants, the dates (including the year!) and places of collection, and any other information that might be helpful. It is essential to write in pencil rather than ink—pencil is more reliably persistent, and advisable not to use typewritten stick-on labels. These have developed the trick of becoming unstuck-on labels to an almost magical degree.

Those trees that produce small dry seeds can be harvested and their seeds treated like any other. A few genera, amongst them *Acer*, *Fraxinus* and *Tilia*, however, include species whose seeds are not very tolerant to being dried and stored.

Collecting Dry Seeds in Fleshy Fruits

Succulent fruits are produced as a lure to entice animals to eat them and distribute the seeds which they contain far and wide across the countryside. This is particularly necessary with long-lived plants like shrubs, which would provide overwhelming competition to seedlings attempting to emerge from within the shadows of their parents. More often than not, in fact almost always, the seeds themselves are not succulent. Once free from the fleshy tissues of the fruit they are revealed as small, hard, dry and often brown objects, very similar to the seeds produced so abundantly by capsules and follicles and other dry fruits. Their appearance is not misleading. The seed of dry and succulent fruits can almost always be treated in exactly similar ways, in particular, in the ways that they are stored. They need to be collected, and treated afterwards, in such a way as to make allowance for the presence of the succulent, fleshy tissues, and provide simple ways to get rid of them, without having to rely on the digestive processes of birds or mammals.

Shrubs that Produce Fleshy Fruits with Dry Seeds

Actinidia	*Fuchsia*	*Rhamnus*
Akebia	*Hedera*	*Rosa*
Amelanchier	*Hippophae*	*Rubus*
Aralia	*Ilex*	*Sambucus*
Berberis	*Ligustrum*	*Sarcococca*
Callicarpa	*Lonicera*	*Schisandra*
Chaenomeles	*Mahonia*	*Skimmia*
Cornus	*Nandina*	*Stachyurus*
Cotoneaster	*Parthenocissus*	*Symplocos*
Daphne	*Pernettya*	*Vaccinium*
Decaisnea	*Poncirus*	*Viburnum*
Desfontainea	*Prunus*	*Viscum*
Elaeagnus	*Pyracantha*	*Vitis.*
Fatsia	*Raphiolepis*	

Trees that Produce Fleshy Fruits with Dry Seeds

Amelanchier	*Ilex*	*Morus*
Arbutus	*Laurus*	*Pyrus*
Cotoneaster	*Malus*	*Pittosporum*
Crataegus	*Mespilus*	*Prunus.*

The best time to collect these fleshy fruits is as they become completely, or almost completely, ripe. The riper and softer the tissues the easier it will be to get rid of them later; but some berries, and daphnes provide a familiar example, are so attractive to birds, that most of them will have been taken long before they ripen. So if the crop is small, or known to be particularly alluring, it will have to be watched daily, and as soon as the fruits start to change colour—probably from green to yellow—collected and put safely out of the way. By this time the seeds they contain will be fully developed and the natural processes of ripening will normally be completed during the following few days. Tissues of fruits often become tacky or sticky, like shrivelled prunes, if they are allowed to lose water after they have been picked, and when they are being harvested they should be put into polythene bags, or some other container which will prevent them from dehydrating. They must not be left too long once fully ripe; if kept lying around for any length of time they quickly start to become rotten, which could later result in the seeds themselves being killed by moulds.

The easiest and best way, to deal with small to moderate quantities of fruits containing small numbers of relatively large seeds, is by hand. The berries and fruits of daphnes, aucubas, skimmias, roses, poncirus, Japanese quinces and decaisnea amongst many others, can all be cut open, and the seeds extracted and dried in the open air, laid out on sheets of kitchen paper in a dry, airy room.

This is not so easy with some berries whose soft tissues enclose numbers of small seeds which are awkward and messy to attack by hand. These can be cleaned mechanically, using a liquidiser or food blender. The berries with the seeds inside them should be mixed with at least five times their volume of water and poured into the blender. This must be run gently at its lowest speed, in short bursts of a few seconds at a time, until the skins and tissues of the berries have completely disintegrated. The addition of more water produces a thin suspension of the fragments of the fruit and the seeds. Mature, ripe seeds are dense enough to sink rapidly to the bottom; the other tissues, including partially developed seeds, either float to the surface or remain in suspension. Once the seeds have settled, the suspension, full of unwanted bits and pieces, can be poured off. If more water is added and the process is repeated two or three times an almost completely clean sample of seeds will be left at the bottom of the blender. The process may sound a little rough, and perhaps too drastic to use on the fruits of rare or precious things: a tomato is a berry, whatever gardeners may call it: use one as guinea-pig.

The seeds are then easy to swill out into a nylon mesh kitchen sieve, and can then be tipped on to a pad of blotting paper, or good quality kitchen paper to dry in a well-ventilated, warm room. Soft paper tissues should be avoided, the seeds may not be entirely free from residues of natural gums, and these can stick to them. Once they have been separated from the surrounding tissues of the fruits these seeds can be treated in exactly the same ways as the small dry seeds obtained from dry fruits; packeted into envelopes and stored in a cool, dry place.

The Large, Moist Seeds of Trees

One of the most familiar of all trees in parks and gardens in temperate parts of the world is the horse chestnut; a plant whose natural distribution is restricted to a few wooded areas in the Balkan Peninsula. It has magnificent flowers and noble leaves, but is most famous for its seeds. The lustrous brown smoothness of conkers is immediately attractive and compels children to collect as many as they can gather. Most are put on one side and a few weeks later they may be rediscovered. Their shine has gone, and they have aged to a dull chocolate brown, their surfaces are creased and folded as they have shrunk in on themselves, and the white flesh of the nut is leathery, discoloured and yellowish.

The conkers that the chestnut produces are nuts, typical of many other nuts and similar fleshy seeds like acorns. They are large, have a high water content and, under dry conditions, lose water through their seed coats. They then shrivel, and the embryos inside them die.

Trees that Produce Large, Moist, Short-lived Seeds

Acer*	Corylus	Juglans
Aesculus	Davidia	Magnolia
Castanea	Fagus	Quercus.

* Some species.

Large moist seeds cannot therefore be kept alive for long periods using the simple conditions which are described later for small dry seeds, and there is no satisfactory method of storing them for periods much longer than a year. Unless they are carefully collected and immediately provided with suitable conditions, many of these nuts are very short-lived indeed: kept in a warm, dry room they will shrivel and be dead within a few weeks.

As soon as they have been collected they must be mixed with barely moist peat, moss or shredded bark and packed into polythene bags in a cool frost-free place—a cellar, if available, an unheated room or a corner of a well-built garage. The salad compartment of a refrigerator provides the right conditions, if there is sufficient space, but holds only small quantities of these large seeds.

Long-term Storage of Small, Dry Seeds

The seed coats of most seeds are permeable to water, but within them is a membrane, which controls the passage of water vapour and gases from the outside world to the interior of the seed. This membrane becomes fully developed only during the last few days as seeds ripen and mature on the parent plants. Seeds which are collected prematurely may have membranes which are not fully developed and this has a lasting, and irredeemable, effect on their ability to perform effectively. Provided seeds have been left to complete their ripening processes they will take up water, in a controlled way, after they are sown. What is not always realised is that seeds in a paper packet in a drawer also absorb water vapour from the air around them. On a thundery, humid day in the middle of summer the water content of their tissues may rise to nearly 20 per cent of their total weight. On a cold, dry frosty January morning the seeds will dry out as the air does, and their water content may drop to as little as 5 per cent.

The frequently repeated fluctuations in water content which seeds experience do them a great deal of harm, and produce stresses which reduce the length of time that they stay alive. Periods when water levels are high are particularly debilitating, and, if seeds remain damp for more than a very short time, they are likely to deteriorate badly, very probably becoming infected with moulds. The appearance of even minute traces of moulds amongst a batch of seeds almost always means that the seeds are dead. This can be used as a rapid, but fairly crude, test to find out whether any sample of seed, maybe put away and half forgotten in drawers and boxes, is worth the trouble of sowing.

Any transparent box can be used to hold the seeds while they are being tested, and a sandwich box with a close-fitting lid is ideal. A pad of wet face tissue should be folded to cover the bottom of the box, and small pinches of all the seeds to be tested, are set out in clusters on the surface of the tissue, keeping each separated by at least 1 cm ($\frac{1}{2}$ in), and making a plan to keep track of which seeds are in each cluster. The lid is then replaced and the box put in a warm place, capable of maintaining a temperature of 20–25° C (68–77°F). After a week the seeds should be examined. Any clusters which have become infested with moulds can be assumed to be dead. There may be others which have started to germinate and are obviously alive, and some will neither have germinated nor gone mouldy. Samples of these should be taken from each cluster and rolled lightly between finger and thumb; if they do not disintegrate they are most probably still alive.

However, such tests can be made almost superfluous by storing seeds in conditions which slow down the rate at which they spoil. Chemical and biological activities proceed faster at high than at low temperatures, and the rate at which seeds deteriorate is no exception. The rule therefore is:

■ the colder and drier seeds are the longer they will survive.

Excellent conditions for long-term seed storage can be provided with very little trouble or expense using equipment available in most households. Low temperatures are the least problem, provided space can be found in a refrigerator or freezer cabinet. Both will prolong life very effectively, although the latter, usually operating at around −13°C (9°F), is very much the more effective.

Keeping seeds dry need be no problem. In this case 'dry' means a water content of around 5 or 7 percent of the total weight of the seed, which is not something anyone can estimate just by looking. Fortunately, there is no need to attempt any weighings or calculations to achieve this. Chemical compounds, known as desiccants, are available which absorb water from the atmosphere and, in a closed container, quite literally dry out the air contained within it, and anything else that's there—such as seeds. One of these desiccants is an innocuous mineral called silica gel. It is easily obtainable, and when put into a closed container with seeds will reduce their water content to the level required.

A simple, inexpensive seed storage kit can be made from a plastic picnic box, a biscuit tin, or any other container which is a suitable size and has a close-fitting, air-tight lid. A layer of silica gel crystals about 1 cm ($\frac{1}{2}$ in) deep is poured in to cover the bottom of the container and the seeds in their paper packets are packed above the desiccant, before replacing the lid. The container can then be stored in the refrigerator or freezer. When the time comes to sow some of the seeds inside it, the box is removed and opened; samples of the seeds to be sown are taken out, and unsown seeds put back in the box which is returned to cold storage.

This simple arrangement provides the twin needs of low temperatures and a

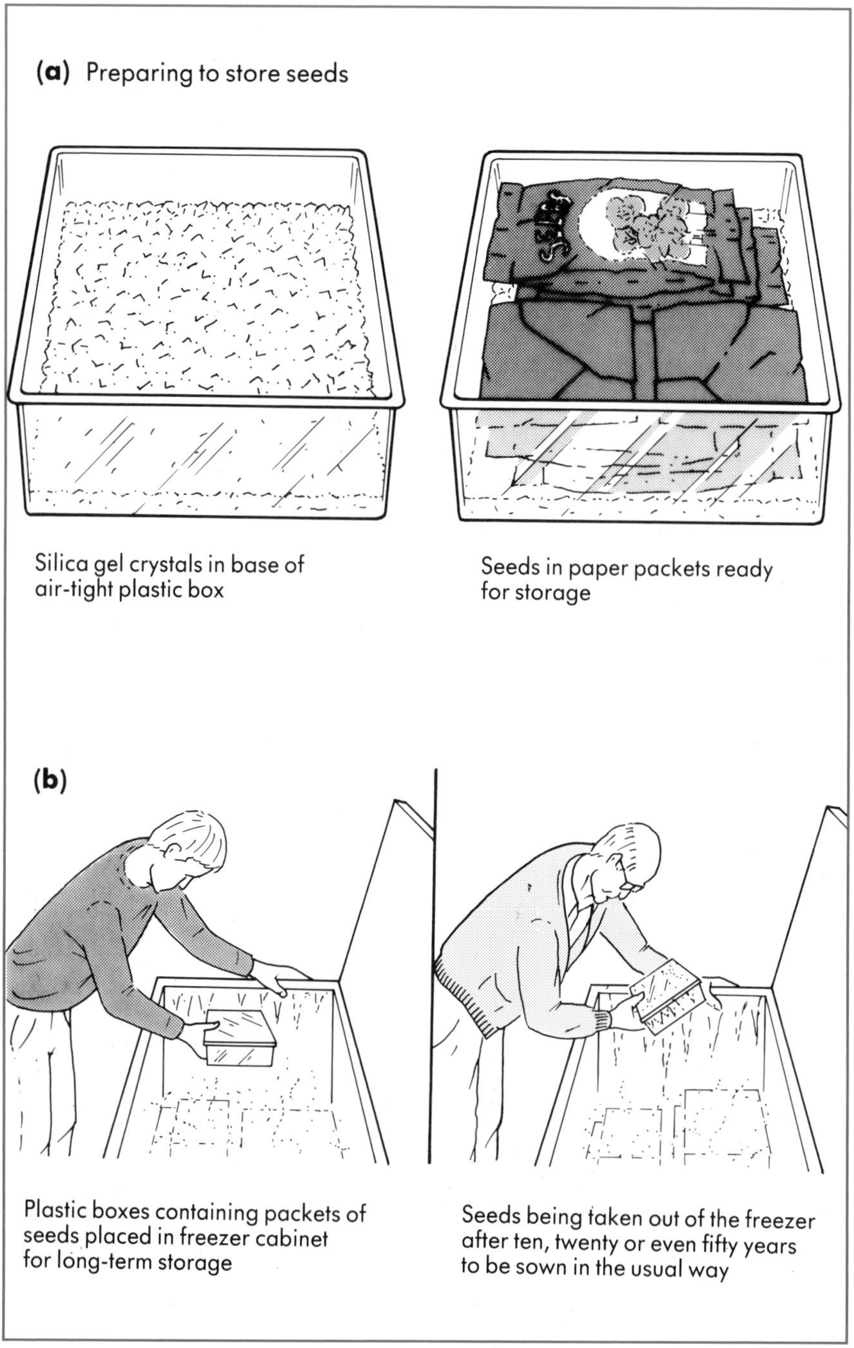

(a) Preparing to store seeds

Silica gel crystals in base of
air-tight plastic box

Seeds in paper packets ready
for storage

(b)

Plastic boxes containing packets of
seeds placed in freezer cabinet
for long-term storage

Seeds being taken out of the freezer
after ten, twenty or even fifty years
to be sown in the usual way

Figure 12.2   Most plants produce small, dry seeds which can be kept alive for very many years using simple and readily available methods of storage.

dry atmosphere, produces very few problems and is not complicated to manage. Silica gel is safe, and its presence in a refrigerator can do no harm. When the time comes to sow seeds from refrigerator they are just taken out of their packets and sown, but, from a freezer, it is preferable to remove the box containing packets of seed about an hour before it is opened. This allows time for its contents to warm up to the temperature of the surrounding air, so that there is no risk of condensation forming on the cold surface of seeds just removed from their store. It is a sensible precaution not to leave boxes of seeds lying open for longer than is necessary, because it will reduce the life of the silica gel, but periods of 10–15 minutes will do no harm. If for one reason or another the boxes have to be taken out of cold storage for a time, even for several weeks every now and then, the seeds will come to little or no harm, provided the lids are kept on to retain the dry atmosphere within them, and they are kept in as cool a place as possible. The eventual life of the seeds will be reduced, but to such a small extent as to be imperceptible.

Sooner or later the crystals of silica gel become saturated with water from the seeds, and from atmospheric humidity when the boxes are opened. When this happens coloured indicator crystals mixed with the silica gel change from blue to pink. These crystals will probably change colour temporarily when the boxes are opened to extract seed, but this can be ignored provided they regain their blue colour after the boxes are closed again. When crystals remain pink the silica gel should be removed and poured into a shallow baking dish to be dried out in an oven. A temperature of about 150°C (300°F) for an hour dries off the water, and the reactivated desiccant can then be returned to the storage boxes, the seeds replaced above them, and the whole lot put back in the refrigerator or freezer.

All small dry seeds can be stored for long periods using this method. It is not a matter of simply keeping seeds alive from one harvest to the next, but of keeping them alive for longer periods of time. Seeds of many attractive plants like the blue poppies, some of the alpine delphiniums, androsace, primulas from the high Alps of Europe and even poppies can be dead within months kept uncaringly in the potting shed. They will survive for several years in an atmosphere kept dry by silica gel in a box in the corner of the study; for ten to twenty years if space for that same box can be found in the refrigerator, and for a hundred years or more in the dark, undisturbed recesses of the freezer.

While seeds are being stored at low temperatures the conditions they need in order to germinate will not change, or will change scarcely at all; stored seed should almost always be treated in exactly the same way as good quality, unstored seed. The seeds of alpines which need low temperatures to reassure them that winter has passed before they will germinate; shrubs that need stratifying treatments or Peruvian lilies that depend on high temperatures followed by mechanical damage to their seed coats will still need these treatments, even though their dry seeds might have spent the previous 20 years deep frozen in the freezer. When they are removed from store they must be sown, allowed to imbibe water, and then given the conditions they need before they are able to produce seedlings.

CHAPTER THIRTEEN

Year-round Propagation

Let us suppose that we have a house with a largish garden, say 1350 sq m (⅓ acre) or more, containing a variety of quite interesting plants, but neglected and run down. Now we have a bit of time to spend on it, and, most rashly, have embarked on a complete overhaul and renewal. Let us also suppose that this garden contains a small greenhouse measuring about 9 sq m (11 sq yd), and that we are prepared to make a special effort, for a year or so, using this to propagate the plants we will need to replant the garden. What should we do to make the most of what we have? And, what might we hope to be able to do?

Fitting Out the Greenhouse for Propagation

First, the greenhouse must be cleared out, broken vents and smashed panes of glass replaced, and thoroughly cleaned up; if it can be left empty for a month or two afterwards to starve any remaining whitefly to death—so much the better. The vents should be counted: more often than not a house this size will have one in the roof and no more. If so, a second roof vent must be fitted on the side opposite the existing one, and at least one, preferably two louvred vents put in the side walls at bench level. The roof vents should be provided with a mechanical system which opens them automatically when conditions become too warm, and closes them again when it cools down.

The house should be fitted out with benches, down both sides and across the end opposite the door, with just sufficient room to move, and work comfortably, down the path between them. These benches are the working surfaces on which the plants being produced will be grown, and they should be laid out to make sure that every centimetre of available space is used, and constructed so that they are strong enough to support the weight of a complete cover of pots and trays full of wet compost and plants. One of the less inspiring starts to a gardener's day is to emerge from breakfast to find that a greenhouse bench has collapsed, and the first hour or two must be spent salvaging what can be retrieved from the wreckage. Ready-made benches constructed from aluminium or slatted wood can be bought—at a price. These seldom make good use of the space available: they are usually too narrow and fail to fill the corners and ends of the house; often they provide barely two-thirds of the bench area that the house is capable of holding. If the benches are fitted into a 9 sq m (11 sq yd) greenhouse, so that no space is wasted they should provide a surface of just over 6 sq m (7 sq yd) on which to grow plants; if they don't, then

184

the path is too long or too broad, or the benches are too narrow. Fitted benches can be made without difficulty, using only rough and ready carpentry, from second-hand timber liberally soaked in a wood preservative. If creosote is used it will preserve the timber, and destroy every plant put in the house for years to come; but other preservatives are available in forms which will do no harm to plants, and preserve the timber at least as effectively.

There must be a solid path, preferably made of paving slabs, down the centre, between the benches; and at least part of the bench area should be heated, to overwinter plants and to make it easier to persuade seeds to germinate and cuttings to produce roots. The most economical and flexible way to do this is with soil-heating cables controlled through a thermostat with variable settings from about 5–25°C (41–77°F) installed as described earlier (p. 36). In this case the simplest arrangement is to use one of the side benches as the heated area, and this will provide about 1.75 sq m (2 sq yd) of heated bench space. A soil-heating cable 25 m (27 yd) long and rated at 300 watts would be enough to keep this area warm if the cables are laid in about 10 cm (4 in) apart. It would be still better to have the entire surface of the benches heated with cables, installed as three independent units each made from a 25 m (27 yd) cable with its own thermostat. The initial cost would be well repaid; by making the surface of the greenhouse more flexible, by increasing its capacity, and by reducing time spent on moving plants around. This heated area is really the heart of the whole system, and arrangements must be made to use the greenhouse space economically, and to provide sufficient additional space in cold frames and nursery beds to hold all the plants that can be produced from it in the course of a year. One problem of being all electric is the looming spectre of power cuts. Something must be kept as a reserve, and the best answer is a paraffin heater, backed up by sheets of bubble polythene to lay over plants during cold spells and in an emergency.

Providing Water

Plants in the greenhouse will need to be watered, and this should be done from a watering can, by hand, using a metal can with a long spout and a good quality, non-drip rose. This is one of the most important pieces of equipment in the garden, and well worth spending a little money on. Automatic systems, or spray lines alike, are much too unreliable, haphazard and unselective to be thought of for a propagation house, and rapid forays, splashing around with water pouring out of a hosepipe, will do more harm than good except during heatwaves. But, once out of the greenhouse and into the frames, a hosepipe becomes the only way of coping with the increased needs of the plants.

Watering is to be done, and a supply of water must be provided, and, in greenhouses, this seems to lead inevitably to thoughts of rainwater collected on the roof and stored in water butts. The best place to put rain water is into a well-made soak-away or the storm drains; water butts, at best, have an inconvenient habit of running out of water whenever it is needed most, and, at worst, harbour various fungi which cause damping off diseases, black-leg or worse. If the desire for a water butt becomes so overwhelming that it cannot be resisted, it must be sited outside the greenhouse—inside it wastes space which could be used to grow precious plants. The best source of water is a stand pipe connected to the domestic supply, and sited immediately outside the greenhouse. This may cause problems in winter when cold weather could freeze it solid, but these can be avoided by wrapping it up well. A stand pipe inside is hard to site without losing some space, especially when cans must be filled and manoeuvred away from it, and, when it is attached to a hose pipe to water other plants, or wash down the car, there is a risk, which should really be called a certainty, that water, spraying wildly from bad connections, will harm cuttings and seedlings at critical stages of their development.

Frames and Nursery Beds

Once the greenhouse has been attended to, thoughts should turn to the frames which will be needed to shelter plants removed from its protection, and to accommodate rooted cuttings, germinating seeds and newly made divisions at almost any time of the year. These should be in a sheltered, sunny position as close as possible to the greenhouse, and linked to it by a paved path. The simplest and most economical way to make them is from old railway sleepers, baulks of second-hand timber or precast concrete building blocks, with standard dutch light frames, covered with a double skin of heavy gauge polythene in place of glass. The area they cover should be equivalent to at least twice the bench space occupied by plants in the greenhouse. A dutch light covers exactly 1 sq m (1 sq yd) and 15 will be needed to provide adequate accommodation for the plants being produced to fill our new garden.

Propagating and looking after plants takes up quite a bit of time, and we want to make full use of the greenhouse and cold frames to produce as many plants as we can, as quickly as possible. Time spent watering, feeding and weeding young plants in their containers as they grow big enough to plant in the garden is all time lost, which could be better spent sowing seeds, taking cuttings or dividing plants. To avoid wasting time in this way, nursery beds will be used to hold young plants whenever possible, as soon as they are large enough, and the weather fair enough, for them to be planted outside. Narrow raised beds should be constructed to make a small nursery (p. 31) for young plants somewhere close to the greenhouse and frames and within hose-reach of a water supply. These must provide sufficient space to grow on seedlings and cuttings until they are large enough to look after themselves when planted out in their permanent positions in the garden; in this case an area of about 200 sq m (240 sq yd) will be needed. A 3 cm (1¼ in) thick mulch of peat or bark worked in by hand between the plants will be used to reduce watering and weeding after they have been set out in their rows in the beds.

There are figures buried amongst the descriptions just given, which provide a guide to the amounts of space needed to shelter and accommodate plants which are being propagated and grown on before being planted in the garden. The numbers required may be greater or smaller—and the garden we are thinking of stocking at the moment is a largish one—but the relative sizes of greenhouses and cold frames, and the amount of nursery space needed will vary, more or less, in the proportions suggested here. In very general terms an effective system for propagating and growing on plants will be provided if:

■ At least a quarter to one-third of the bench space in a greenhouse is provided with soil-heating cables.

- Cold frames cover an area two to three times that of the benches in the greenhouse.
- Nursery beds are at least ten and preferably 15 times the area of the usable space in the cold frames.

Getting Going

The greenhouse is set up, frames are provided for, and space for nursery beds is available, although the latter will not be needed for a while and can be constructed one by one as plants to fill them become available. All is ready to begin propagating plants, and, as it happens it is 1st June, which is a very convenient moment to start. The spring planting rush is over, the debris and weediness left by winter have been cleared away and the plants in the garden are beginning to grow, and produce the shoots and seeds which will be used to propagate more. The following calendar gives a guide to what might be done month by month using the accommodation and facilities described above. It is only fair to add that to achieve what is described would need dedication and effort well beyond the normal call of gardening duty, and that good fortune would have to bless the endeavour with unusually high proportions of success when striking cuttings, or persuading seeds to germinate. Mere mortals should be content with less.

Initial capital letters are used in each paragraph to identify particular batches of plants as they make their way through the different processes involved in raising them from cuttings, seeds or divisions until they take their places in the garden or the nursery, as follows:

A. Cuttings from the soft, immature shoots of shrubs.
B. Juniper cuttings.
C. Basal cuttings from herbaceous perennials and alpines.
D. Summer cuttings from the semi-mature shoots of shrubs.
E. Seeds of alpines, perennials, shrubs and bulbs.
F. Cuttings of heathers and heaths.
G. Cuttings from ericaceous shrubs, like Japanese azaleas.
H. Divisions of herbaceous perennials.
I. Cuttings of tender perennials.
J. Cuttings of silver-leaved and semi-tender shrubs.
K. Conifer cuttings.
L. Hardwood cuttings of deciduous shrubs and roses.
M. Seeds of alpines, perennials and shrubs.
N. Seeds of annual bedding plants.

A Propagation Calendar

JUNE

A. Soft tip cuttings of shrubs (p. 128) will be ready, and can be set up in 9 cm (3½ in) square plastic pots, in a compost made up of two-thirds horticultural grit and one-third vermiculite or calcined clay*. There is room for 16 pots with 16 small cuttings in each.

B. By mid-June and into July, juniper cuttings will be ready (p.68). Set up in 13 cm (5 in) square plastic pots in the same cutting compost. Fifteen pots, each containing 15 cuttings.

C. Herbaceous plants (p. 110) and alpines (p. 92) are producing basal shoots which can be used as cuttings. Space can be found for ten pots each 13 cm (5 in) square and holding 15–25 cuttings in each, using the same cutting compost.

All these will benefit from a humid atmosphere, enclosed within a polythene cover, and temperatures of 15°C (59°F) at their roots.

*Here and elsewhere the term 'calcined clay' is used to describe a variety of materials including arcillite, cat litters, and materials marketed to mop up oil or similar spillages from garage floors (p. 56).

JULY

A. Soft tip cuttings should be potted up, as they produce roots, into individual 7 cm (3 in) square plastic containers, filled with a standard loam- or peat-based compost. Set out on the unheated greenhouse bench to grow on.

C. Basal cuttings potted up individually as each batch produces roots into 9 cm (3½ in) square pots (herbaceous) and 7 cm (3 in) square pots (alpines) containing standard potting compost. Grow on on the unheated part of the greenhouse bench.

D. Semi-mature shoots of some shrubs will be ready to use as cuttings (p. 130). Room for 25 pots, each 13 cm (5 in) square and holding up to 25 cuttings in each in a grit/vermiculite compost. Set up under polythene cover with minimum temperature of 15°C (59°F).

E. Towards the end of the month seeds can be sown, which will germinate without special treatments, of alpines (p. 88), shrubs (p. 122) and herbaceous perennials (p. 104), above soil-heating cables at 15°C (59°F). These seeds can be sown until the end of August in 9 cm (3½ in) square pots using methods described previously (p. 56).

F. Cuttings of heaths (p. 71) can be taken during July and throughout August, and set up at 15°C (59°F), under cover. Space for up to 40 square plastic pots, each of 13 cm (5 in) and holding up to 64 cuttings in a cutting compost made up of equal parts of grit and vermiculite.

G. Cuttings of Japanese azaleas and other ericaceous shrubs can be set up at the same time and in the same conditions as the heaths. Space for up to ten pots, 13 cm (5 in) square, each holding 25 cuttings.

H. Towards the end of the month and through August herbaceous perennials (p. 119), including petaloid moncotyledons (p. 165) growing in the garden, can be dug up and divided. Divisions should be potted straight into 7 cm (3 in) square pots using a standard loam- or peat-based compost. Space for 20 pots of each of 40 different varieties, set out in frames with lights closed, and shaded, for the first fortnight.

AUGUST

A. Soft tip cuttings should be well-established in their pots, and can be moved into a cold frame; after a month they should be ready to line out in the nursery or even plant out in the garden—semi-tender varieties should be kept in the cold frame till the following spring.

B. Juniper cuttings will have formed roots, and should be potted up individually in 7 cm (3 in) square pots, using a peat-based compost, replaced on unheated bench in greenhouse and moved out into the cold frame after a fortnight.

C. Basal cuttings of herbaceous plants and alpines moved to cold frame; ready to plant out in garden during September and October.

D. Semi-mature cuttings of shrubs which are well-rooted can be potted up into 7 cm (3 in) square plastic pots, preferably into a loam-based compost, and moved on to an unheated bench in the greenhouse; after about a month they can be transferred to the frame.

E. A few of the most advanced seedling perennials, which germinate rapidly and grow large quickly, should be potted individually into 7 cm (3 in) square pots containing a loam-based compost, and placed on the unheated greenhouse bench for a fortnight. They can then be transferred to the cold frame to overwinter.

SEPTEMBER

D. Cuttings from semi-mature shoots of shrubs which produce roots after the beginning of September should not be potted up (p. 81), but moved to an unheated part of the greenhouse, and given a drench with liquid feed (p. 81), and then fed repeatedly at fortnightly intervals till they lose their leaves.

E. Remaining seedlings of perennials, alpines and shrubs should have germinated and produced small plants. These can be moved to the unheated benches in the greenhouse. Any pots which contain no seedlings should be moved into frames to overwinter; seedlings may appear the following spring.

F. Cuttings taken from heaths should be forming roots; as they do so they can be transferred to the unheated benches in the greenhouse to overwinter.

G. Cuttings of Japanese azaleas can be treated in the same way as the heaths. On no account should either be potted up individually before the winter. If space in the greenhouse is needed during the winter, these and the heaths can be moved into the frames.

H. Divisions of herbaceous perennials can either be overwintered in the frame, or planted out in the nursery or in their final positions in the garden.

I. Cuttings of tender perennials (*Gazania, Pelargonium, Calceolaria,* Heliotrope, etc.) can be taken, and set up in 13 cm (5 in) square pots with 9–15 in each, using a gritty cutting compost: 50:50 grit and calcined clay or vermiculite. Space for up to 15 pots, to overwinter on heated* bench in greenhouse.

J. Cuttings of silver-leaved, aromatic and semi-tender shrubs (p. 140) can be taken now; set up in 13 cm (5 in) square pots each containing 15–25 cuttings, in a 50:50 grit and vermiculite compost. Space for up to 60 pots to overwinter on the heated* bench.

* Note that from early November until the beginning of March the thermostat controlling the temperature of the heated part of the greenhouse bench should be turned down to 5°C (41°F). Sheets of bubble polythene should be cut to convenient sizes and used to cover up all the plants on frosty nights, and by day during prolonged frosty spells.

OCTOBER

D. Cuttings of semi-mature shoots of deciduous shrubs can be overwintered in the greenhouse, or if space is insufficient moved out into the cold frame.

K. Cuttings of all kinds of conifers (p. 73), apart from junipers, can be taken through October and into November, using 13 cm (5 in) square plastic pots with 25 cuttings in each, in a 50:50 grit and vermiculite cutting compost. Set up on heated bench in greenhouse to overwinter; space for 15 pots.

L. Hardwood cuttings of shrubs (p. 133) and trees (p. 152) should be taken towards the end of the month as the leaves fall, and set up in 3 litre (5 pint) polypropylene pots in the cold frame in a 3:1 mixture of grit and vermiculite. Fifteen to 20 cuttings in each pot, and space for 20 pots.

NOVEMBER

D. Overwintering rooted cuttings of semi-mature shoots of deciduous shrubs will have lost their leaves, and should have dead leaves removed, and any mosses or liverworts on the surface of the compost taken off. These cuttings should now be kept as dry as possible, without becoming desiccated, until early March.

I. Cuttings of tender perennials will have produced roots, but should not be fed, and should be kept very dry until March. When absolutely necessary they should be watered, but only during the morning on bright sunny days. They can be watered by partially immersing the pots in a deep tray of water, and leaving them to stand and absorb water from below for 10–15 minutes.

J. Cuttings of silver-leaved plants, etc. should be treated in the same way as tender perennials. Both sets must be well-protected by covering them with a sheet of bubble polythene, laid over the tops of the plants, on cold nights. In very cold weather the setting of the thermostat controlling the temperature of the compost may have to be increased to 10°C (50°F) to keep it frost-free.

DECEMBER

No new propagation started. Maintenance only; all plants whether in the greenhouse or in the cold frame must be watered very sparingly, and conditions must be avoided which result in the leaves of plants remaining wet after nightfall.

Plants overwintering in the greenhouse should be given as much ventilation as possible every day, though ventilators should be closed during the nights. The lights above plants in the cold frames should be wedged open continuously except when temperatures fall to freezing point, or winds are exceptionally cold or drying.

JANUARY

M. Seeds sown of alpines (p. 87), perennials (p. 103) and shrubs (p. 122) which depend on low temperature conditioning before they will produce seedlings. Sown in 13 cm (5 in) square pots and placed in cold frame to be conditioned (p. 87) by natural low temperatures during the late winter and early spring. Space and facilities for up to 20 pots.

An excellent time to take a break from propagating, gardening and everything to do with plants and go on holiday. Preferences might be for a skiing holiday or a trip to the Nile Valley or some other place entirely free of visible vegetation.

FEBRUARY

Maintenance continued as for December. Spells of warm sunshine at this time of the year can dry out plants in the greenhouse very rapidly, and may catch unwary gardeners out.

D. All rooted cuttings of semi-mature shrubs still in the greenhouse moved into frames for the rest of the winter, and until they come into growth naturally in the spring. Given a liquid feed towards the end of the month.

MARCH

E. Summer/autumn sown seedlings of perennials, alpines and shrubs removed from their pots and potted up individually into 7 cm (3 in) square plastic pots containing standard loam- or peat-based compost. Placed in a cold frame, and lights kept closed down above them for a fortnight or three weeks until they start to grow.

F. Rooted heather cuttings moved into frame, if not already there, and fed with liquid feed. Full ventilation maintained whenever possible.

G. Cuttings of Japanese azaleas treated in the same way as the heather cuttings.

I. Rooted cuttings of tender perennials fed, and potted individually into 9 cm ($3\frac{1}{2}$ in) square pots using a peat-based compost. Set out on unheated bench space in greenhouse.

J. Rooted cuttings of silver-leaved, aromatic and semi-tender shrubs potted into 7 cm (3 in) square plastic pots, using a peat-based compost. Set out on unheated bench space in greenhouse.

N. First sowings of annual bedding plants made (p. 48), using 9 cm (3 in) square plastic pots. Early sowings should be confined to hardy species of mediterranean origin (p. 46).

APRIL

A. Overwintered plants raised from tip cuttings removed from cold frame and planted out in garden.

B. Cuttings of junipers removed from cold frame and lined out in nursery.

F. As soon as heather cuttings show signs of growth they can be potted up individually into 7 cm (3 in) square pots in an ericaceous peat-based compost, and returned to the cold frame to grow on.

J. Plants of silver-leaved, aromatic and semi-tender shrubs moved into cold frame, ventilated whenever possible during the day, but lights closed during the nights.

L. Hardwood cuttings examined, and any which have formed roots potted into 9 cm (3½ in) square pots using either a loam- or peat-based potting compost, and returned to frame.

M. Seedlings of alpines and perennials sown during January transferred to greenhouse to germinate; any which have done so should be pricked out into half trays.

N. Annuals pricked out as they are ready into seed trays, and kept under cover in the greenhouse. More sowings made, up to a maximum of 100 different kinds.

MAY

D. Plants raised early from cuttings of semi-mature shoots and potted up the previous summer lined out in nursery bed to grow on. Remainder potted up individually in 9 cm (3½ in) square pots and returned to frame. These should be ready to line out in the nursery a month to six weeks later.

E. Seedling plants of perennials, alpines and shrubs can be lined out in the nursery beds as they become large enough through May and June.

G. Rooted cuttings of Japanese azaleas potted up individually in 7 cm (3 in) square plastic pots using a peat-based ericaceous compost, and returned to cold frame. Ventilators opened through the day, but kept closed at night.

I. Cuttings of tender perennials moved into cold frame. Ventilated when possible except during cold days, and at night.

J. Young plants of silver-leaved, aromatic and semi-tender shrubs lined out in nursery beds to grow on or planted out in the garden.

K. Rooted conifer cuttings potted individually into 9 cm (3½ in) square pots containing a peat-based compost as soon as possible.

N. Trays of bedding plants transferred to cold frame. Hardy plants grown from early sowings planted out in the garden.

JUNE

F. Heather cuttings lined out in nursery beds towards end of month to grow on.

G. Japanese azaleas to grow on in cold frame; larger ones ready to transfer to nursery bed towards the end of the month and the remainder in July.

I. Tender perennials planted in the garden.

K. Conifer plants grown on in cold frame, fed towards end of month with liquid feed, and planted out in nursery bed during July,

M. Seedling plants of alpines, perennials and shrubs sorted through; more vigorous varieties lined out in nursery beds, the remainder kept in their pots, and grown on sheltered by sides of cold frame with the lights removed.

N. Remaining bedding plants set out in the garden.

A glance through the schedule outlined above leaves little room for doubt that it would be a taxing one to undertake: it also vividly illustrates what a powerful tool a greenhouse can be, provided that a part of it can be heated, and the frost can be kept out of it through the winter. The complete list of the plants that would have been produced if the whole programme ran through smoothly and successfully runs as follows:

      800 Herbaceous perennials from divisions.
    1230 Silver-leaved, aromatic and semi-tender shrubs.
      375 Assorted conifers.
    2000 Seedlings of herbaceous perennials and shrubs.
    2560 Heaths and heathers.
      250 Japanese azaleas.
      700 Seedling perennials, shrubs and alpines.
      300 Shrubs and trees from hardwood cuttings.
      100 Trays of bedding plants/annuals.*
      256 Shrubs from tip cuttings.
      375 Junipers.
      135 Tender perennials; *Pelargonium*, *Gazania*, heliotrope etc.

* A large number of bedding plants are included during this first year to fill up spaces in the garden, and provide something to look at and enjoy while the perennials, shrubs, trees and alpines are growing on in the nursery and frames.

After-care

At the very beginning of this book the suggestion was made that propagation is often regarded as a rather advanced gardening skill and, time and again on a nursery, it will be found that the jobs which involve producing plants by propagating them from cuttings or seeds or grafts are the most sought after, the most highly regarded and amongst the best paid. More often than not the tasks which have to be done later just to grow the plants are less highly thought of, and watering, spacing out and feeding plants, amongst other regular chores, are very frequently left to be done by the most inexperienced staff on the place. A similar viewpoint often finds an echo amongst gardeners who will be delighted to show off the plants they are propagating from cuttings and seeds, to share their successes, and talk about the equipment they have or hope to have, and yet will pass by yesterday's cuttings and yesterday's seedlings, with scarcely a glance to see how they are getting on.

When they are looked at it is very often obvious that they are not getting on. They may be lying about in higgledy-piggledy rows, or crowded together one on top of another; weeds may infest them, and signs of vigorous healthy growth can be very hard to discern. Some will eventually be planted out, some given away, some trotted off to plant stalls, and the remainder will tarry for awhile and then either die or be thrown out.

Propagation is interesting, it is full of variety and challenges, and there are few things in gardening more satisfying than finding a way to propagate a plant, which has thwarted us in the past, and few pleasures greater than being able to go around a garden and know that most of the plants to be seen are home grown.

Nevertheless, the tasks which really demand skill, and an expert intimate knowledge of plants and the ways they behave, are the simple chores of watering, and feeding, and putting a plant in the one spot in a garden which suits it, and where it will grow best. These all depend on understanding the infinite tiny variations in preference which makes one kind of plant different from another, and which under natural conditions allows them to form tangled, complex and amazingly varied communities in which each plant has a place, and from which outsiders are almost completely excluded.

In quite recent years the world of gardening has invented a new title by which to distinguish those who are thought to be experts; they are called plantsmen or plantswomen, and it sometimes seems that they are chiefly notable for the large number of Latin names which they know. Perhaps the skills which we should really look up to are those which gardeners develop through instinct and experience by which they learn how to care for even the most demanding of plants, the positions where they will flourish, and the companions they best mix with and combine with to form subtle and harmonious communities in a garden.

CHAPTER FOURTEEN

Propagation Table

One of the first things to do when planning to propagate plants is to take a good look at the methods available and the times of the year when they can be done. This makes it possible to compose a propagation calendar covering all the plants we would like to produce, showing which methods are most appropriate for the particular circumstances, and avoiding situations where there is more to be done than time or facilities allow for. In the table that follows the genera are listed in alphabetical order of their Latin names. The index provides cross-references to common names and some Latin synonyms when widely used alternatives exist. The appropriate methods of propagation and the months of the year are listed for each genera.

The table is intended as a concise and simple guide, but, in the space available, can be neither comprehensive nor descriptive of methods to be used. The entry for each genus is restricted to a single line: sometimes, for example with genera made up entirely of species of annual plants, this provides adequate information, but the entries for large genera, containing a variety of plants, or from diverse parts of the world, inevitably need more careful interpretation, and should be supported by reference to more complete descriptions in other parts of the book.

Methods of propagation are grouped under the three headings used consistently throughout this book, in the following order:

Seed.
Cuttings.
Divisions.

The months of the year when different methods of propagation can be done are based on the conditions in the southern parts of Britain. They have been deliberately stated in fairly broad terms, and can be applied to most cool temperate parts of the world. Differences from one area to another usually have only a small effect on the time of the year when different gardening operations must be done—often little more than those that occur from one year to another in a particular place. Much more significant as a rule are the variations found from place to place in the intensity of winter cold, summer heat and the distribution of rainfall. These may not make a great deal of difference to the times when different operations are done but can very strongly influence which plants can be grown, the ways that they are grown, and the protection and facilities needed for a successful result. Examples are frequently found of plants for which winter protection or artificial warmth, described as merely helpful in southern Britain, are essential for survival in colder areas. The use of supplementary heating to encourage cuttings to produce roots, which may be beneficial during the summer in Britain, would not just be unnecessary but actually harmful in southern France.

Key to the Propagation Table

Growing Plants from Seeds

Cld Seeds which will germinate without artificial heat. Some of these, sown during the winter, depend on exposure to a period of low temperatures before they are able to produce seedlings. Many of those sown in the spring, or during the summer, can also be successfully grown under warmer conditions.

Htd Seeds which respond to warmer conditions than those which prevail naturally at the seasons when they are usually sown. Most of these will benefit if grown in heated greenhouses, or propagators, in which soil temperatures can be maintained at 15–25°C (59–77°F).

Cnd Seeds which depend on special conditions to ensure that they germinate satisfactorily. Some may be simple to provide; others, more complex, may require successive transfers from one set of conditions to another.

Plants that can be Grown from Cuttings

Tip Cuttings prepared from immature tips of young shoots. These are very vulnerable, and depend on careful treatment under closely controlled conditions.

S/M Semi-mature shoots taken from shrubs during the summer provide a widely applicable, straightforward means of propagation.

Hdw Hardwood cuttings are made from fully mature, woody shoots, usually late in the autumn, during the early winter or just before growth starts in the spring.

Bas Basal cuttings can be prepared from the young shoots of herbaceous plants; these appear at, or from just below, ground level during the spring and early summer.

Rt Root cuttings are prepared from strong roots filled with storage reserves, and provide a means of propagating a variety of trees, shrubs and herbaceous plants.

Bud Single bud cuttings, usually made from axillary buds found in the angles between leaves and stems, can be used to propagate a number of shrubs and climbers, and a few herbaceous perennials.

Methods of Propagation by Division

Div Many plants produce shoots of one kind or another which form roots naturally as they develop. These can be removed from their parents as ready-made plants.

Lay Layers, prepared from the shoots of shrubs and some trees, produce roots while still attached to their parents. Subsequently they are removed and grown on to produce independent plants.

Blb Offsets produced by many bulbous plants are a slow but steady means of increase. Sometimes their rate of production can be increased by appropriate treatments.

The months of the year are numbered consecutively from January to December, and simple conventions are used to indicate time spans as follows:

3/4 during March and April
3–5 from March to May
3 : 7 in March and again in July

Propagation Table

NAME	SEEDS			CUTTINGS						DIVISIONS		
	Cld	Htd	Cnd	Tip	S/M	Hdw	Bas	Rt	Bud	Div	Lay	Blb
Abelia				4/5	7/8							
Abies	11/12											
Abutilon		1–3		5	7/8	10/11						
Acaena		2–4					5–7			5/6		
Acanthus		3–5						12–2				
Acer	9–11		9–11	4/5		9						
Achillea		4/5					6–9			4–6		
Aconitum			7/8							3–5		
Acorus										5/6		
Actaea			10/11							4/5		
Actinidia	10/11				6/7						10/11	
Adenophora	7/8	2–4										
Adiantum		4–6								4/5		
Aegopodium										4–8		
Aesculus	10/11							12/1				
Aethionema	2/3	1–4					6/7					
Agapanthus		3–5								4/5		
Ageratum		3–5										
Ailanthus								11–2				
Ajuga							4–6			5–7		
Akebia	7/8				6/7						10/11	
Alchemilla	1/2									7/8		
Alisma										3/4		
Allium		2–5	7–10									3–5
Alnus			10/11									
Alonsoa		3–5										
Alstroemeria			1/2									
Althaea	7/8	2–6										
Alyssum	7/8	3/4			6/7							
Amaranthus		4/5										
Amelanchier			7/8	5						9/10	10/11	
Ammobium		4/5										
Anaphalis	3/4	1–3					5/6			7/8		
Anchusa	6/7	3/4					4/5	1/2				
Androsace			1–3				6–9			7/8		
Anemone			7/8					12–2		7/8		
Antennaria										4–8		
Anthemis		2–4					4–8					
Anthericum			7/8									11–3
Anthriscus			7–9							3:7		
Antirrhinum	8/9	2–4										

NAME	SEEDS			CUTTINGS						DIVISIONS		
	Cld	Htd	Cnd	Tip	S/M	Hdw	Bas	Rt	Bud	Div	Lay	Blb
Aponogeton										4/5		
Aquilegia	5–7	1–3	7/8									
Arabis	7/8					6/7						
Aralia	9–11	3/4						12–2				
Araucaria		12–3										
Arbutus	3/4					11–2					8/9	
Arctotis		3/5					8/9					
Arenaria										4–8		
Arisaema												4/5
Armeria	3/4						7/8					
Artemisia					8/9		6–9			4–7		
Arum												4/5
Aruncus	1–3									3:7		
Arundinaria										4/5		
Asarum										7/8		
Asphodeline	7/8	3/4										3/4
Aster	7/8	3–5					4/5			3:7		
Astilbe	2/3									3:7		
Astrantia			8/9							7/8		
Athyrium		2–5								4/5		
Aubrieta	7/8						6–8					
Aucuba						9–11						
Azara					7/8	10–12						
Ballota					7/8							
Baptisia	7/8	3–5										
Bellis	6/7									7/8		
Berberis			10–12			9–11						
Bergenia	1/2						5/6			7/8		
Betula			10/11									
Bletilla												4/5
Borago	4/5											
Brachycome	5	4/5										
Brassica		3/4										
Briza		3/4								4/5		
Brodiaea		3/4										7/8
Brunnera										7/8		
Buddleia		3/4				6/7	10/11					
Bupleurum			10/11		8							
Butomus										4/5		
Buxus						9/10						
Calamintha					6/7							
Calendula	3:9	3–5										

NAME	SEEDS			CUTTINGS						DIVISIONS		
	Cld	Htd	Cnd	Tip	S/M	Hdw	Bas	Rt	Bud	Div	Lay	Blb
Calla										4/5		
Callicarpa			11/12		6–8							
Callistemon		2–4		5/6	7/8							
Callistephus		4/5										
Calluna					8/9						3/4	
Calocedrus	10/11				7–9							
Caltha			6/7							5/6		
Calycanthus					7/8						4/5	
Camassia	8/9											9/10
Camellia					6/7				2/3			
Campanula	2:7	3/4					4/5			3:7		
Campsis					7/8			12–2				
Cardiocrinum			9/10									9/10
Carex										4/5		
Carlina		2–5						1/2		3/4		
Carpentaria		3/4				3/4					9/10	
Carpinus			9/10		7/8							
Caryopteris	1/2			3/4	6–8	10/11						
Castanea	10/11											
Catalpa					7/8			11–2				
Catananche	7/8	2–4						2/3				
Ceanothus					6–8	2:10						
Cedrus	3/4	2/3										
Celastrus	11	2–5			7/8			12–2			9/10	
Centaurea	4:8	2–4					3–5			4/5		
Centranthus	4/5											
Cephalaria	7/8	1–4										
Cerastium		3–5					6–8			3–5		
Ceratostigma				5	6–8					4/5		
Cercidiphyllum		2–4			6/7						9/10	
Cercis		2–4										
Cestrum				5	7/8							
Chaenomeles	1/2				6–8			1/2			3/4	
Chamaecyparis	3/4				10/11							
Cheiranthus	5/6						7/8					
Chelone	12/1	2–5								7/8		
Chimonanthus		2/3									7/8	
Chionodoxa	5/6											8–10
Choisya					6–8	10/11						
Chrysanthemum	4:8	3–5					3–5			7/8		
Cimicifuga			9/10							3–5		
Cistus		2–5			6–8	9/10						

NAME	SEEDS			CUTTINGS						DIVISIONS		
	Cld	Htd	Cnd	Tip	S/M	Hdw	Bas	Rt	Bud	Div	Lay	Blb
Clarkia	3/4											
Clematis			10/11	4/5	6–8						3/4	
Cleome		2/3										
Clerodendron								12–2		4/5		
Clethra		2–5			6–8						10/11	
Codonopsis	7/8	2–4					4/5					
Coix		2/3										
Colchicum			6/7									7/8
Colutea		2–4			7/8							
Convallaria												2/3
Convolvulus	4/5	3/4			6–8							
Coreopsis	4/5	1–3								7/8		
Cornus			7/8			11/12					5/6	
Corokia						9–11						
Coronilla		3–5			6–8	10/11						
Cortaderia		3/4								4/5		
Corydalis			9/10									
Corylopsis						7/8					4/5	
Corylus			10/11							2/3	9/10	
Cosmos		3–5										
Cotinus	9/10			5/6							4/5	
Cotoneaster			9/10		7/8	10/11						
Crambe	3/4						4/5	11–2				
Crataegus			10/11									
Crepis	2–4									4/5		
Crinodendron					6/7							
Crinum												3/4
Crocosmia		2–4										3/4
Crocus	6/7											8/9
Cryptomeria	4	2–4				9/10						
Cuphea		3/4		3/4								
Cupressocyparis						2:10						
Cupressus		10/11				2/3						
Curtonus			9/10									5/6
Cyananthus			1–3				5/6					
Cyclamen	3:8											
Cydonia		2/3										
Cynara		1–3										
Cytisus	9/10	3/4			7/8	10/11						
Daboecia					7/8	9/10						
Dahlia		3/4					3/4			4		
Daphne			7–9		6–8						5/6	

NAME	SEEDS			CUTTINGS						DIVISIONS		
	Cld	Htd	Cnd	Tip	S/M	Hdw	Bas	Rt	Bud	Div	Lay	Blb
Davidia			10	6	7/8						4/5	
Decaisnea			10/11									
Delphinium	4:7	1–4					4					
Desfontainea		2–4			7/8	10						
Deutzia		2/3			6–8	10/11						
Dianthus	6/7	2–4					6/7				7/8	
Diascia		2–5					5:9			4		
Dicentra	12/1	3/4						2/3		3/4		
Dictamnus			7/8									
Dierama		3–5										10
Diervillea					7/8	10/11				5/6		
Digitalis	5–7	1–4										
Dimorphotheca	5	3/4		5	6–9							
Dipelta					6–8	10						
Disanthus		2/3			6–8						5:9	
Dodecatheon	7/8	3/4										
Doronicum	7/8	1–4								7/8		
Douglasia							7/8					
Draba	2–4						6/7					
Drimys				4/5		10/11					3/4	
Dryas	9/10				6–8					4/5		
Dryopteris		4/5								4/5		
Eccremocarpus		3/4										
Echinops	4–6	1–3						11–2		3/4		
Echium	3/4	2–4										
Edraianthus		2–4					6–8					
Elaeagnus			7:10		7/8	2:10				2/3	9/10	
Embothrium		2–4						12/1				
Enkianthus		2/3			7–9						3/4	
Epilobium	2–4						4/5					
Epimedium			7/8							7/8		
Eranthis			4/5									2/3
Eremurus	7/8											
Erica		2/3			7–9	2:10					10/11	
Erigeron	4/5	2–4					4/5			7/8		
Erinus	4:7	2/3										
Erodium			10/11				4–6	2/3		7/8		
Eryngium	3/4	1–3	7/8					2/3		7/8		
Erysimum	3–4	2/3			7/8							
Erythronium			8/9									7/8
Escallonia					6–9	10/11						
Eschscholtzia	4:9	3/4										

202

NAME	SEEDS			CUTTINGS						DIVISIONS		
	Cld	Htd	Cnd	Tip	S/M	Hdw	Bas	Rt	Bud	Div	Lay	Blb
Eucalyptus		2–5										
Eucomis												4/5
Eucryphia		3/4			6–8	2/3						8/9	
Euonymus			10/11		6/7	10/11						10/11	
Eupatorium	4/5	2–4									3–5		
Euphorbia	3:7	2/3		5	7/8		4/5						
Euryops					6–8								
Exochorda		2–4		4/5	6–8								
Fabiana				4	7/8								
Fagus	11/12												
Fatshedera					6–8					7–9			
Fatsia		3–6								7–9			
Festuca	7/8	3/4									4/5		
Filipendula	12/1	2/3									7/8		
Foeniculum	4/5	2–4											
Forsythia					6–9	10–12							
Fothergilla					6/7							8/9	
Fragaria	7/8	1–4									3:8		
Fremonto-dendron		3/4				3/4							
Fritillaria	7/8									7/8			
Fuchsia		3/4		4/5	6–9								
Gaillardia	4:7	2–4					4				4:8		
Galanthus	4/5												2/3
Galega	4:7	1–4											
Galtonia		1–4											9/10
Garrya						2:10							
Gaultheria	10/11				7/8						9/10	4/5	
Gazania		2–5					8/9						
Genista		2/3			6/7	10/11							
Gentiana		3/4	12/1				4/5				6/7		
Geranium		1–4					5–9				4:8		
Geum	7/8	1–4									4:8		
Gilia	4:9	3–5											
Ginkgo	9/10												
Gladiolus		2–4											10:4
Glaucium	7/8	2/3											
Gleditschia	3/4												
Godetia	4:9	3–5											
Gomphrena		3–5											
Grevillea					6/7	3/4							
Griselinia					6/7	10/11							

NAME	SEEDS			CUTTINGS						DIVISIONS		
	Cld	Htd	Cnd	Tip	S/M	Hdw	Bas	Rt	Bud	Div	Lay	Blb
Gunnera		2/3								4/5		
Gypsophila	4:9	2–4					4–7					
Haberlea			8/9						6/7			
Halesia			9/10	5							3/4	
Halimiocistus					6–9							
Halimium					6–9							
Hamamelis			7/8								3/4	
Hebe					6–9							
Hedera					6/7	10/11			6–9		4/5	
Helenium							4/5			4:8		
Helianthemum		2–4			7–9							
Helianthus	4/5	2–4								4:8		
Helichrysum		3/4		4/5	6–9							
Helictotrichon										4/5		
Helipterum	4/5	3–5										
Helleborus			6/7							3/4		
Helxine										4–9		
Hemerocallis			9/10							7/8		
Hepatica										7/8		
Heracleum	3/4											
Hesperis	4:7									4:7		
Heuchera	4/5	1–4								4:8		
Hibiscus					7–9							
Hippophae	10/11											
Hoheria		3/4		5/6	8						4/5	
Holodiscus					7–9							
Hordeum	4:8	3/4										
Hosta	6/7	2–4								7/8		
Houttuynia										4:8		
Hyacinthoides	7/8											6/7
Hyacinthus	6/7								7/8			
Hydrangea		3–5			7–9							
Hypericum	1/2				7–9	10/11				3/4		
Hyssopus	4/5	2–4			6–9							
Humulus		4/5					4/5					
Iberis	3–6				6–9							
Ilex			10/11		7/8	2:9					10/11	
Impatiens		4–6		5–9								
Incarvillea	7/8	1–4										
Indigofera		2–5			6–9							
Inula		3/4								3:8		
Ipheion	6/7											6/7

NAME	SEEDS			CUTTINGS						DIVISIONS		
	Cld	Htd	Cnd	Tip	S/M	Hdw	Bas	Rt	Bud	Div	Lay	Blb
Ipomoea		2–4										
Iris		3/4	8/9									6–8
Itea						8/9						
Jasminum					6–8	10/11						
Juglans	10:3											
Juniperus					6/7	4						
Kalmia		3–5				8/9					8/9	
Kerria					6–8	10/11				3/4		
Kniphofia		3/4										3/4
Kochia		3/4										
Koelreuteria		2–4						12–2				
Kolkwitzia				4/5	7–9							
Laburnum	4	3/4				10–12						
Lagurus	8/9	2–4										
Lamiastrum					5–8					3/4		
Lamium	3/4				5–8					3/4		
Larix	2/3											
Lathyrus	7/8	2–4										
Laurus						11/12					9/10	
Lavandula	6–8	2–4		4/5		8/9						
Lavatera	4/5	2–4			6–8	9/10						
Leontopodium	1/2											
Leptospermum					7/8	10/11						
Leucojum	7–9											6–8
Lewisia	3:7		7–10							6/7		
Liatris	4:7	1–4								3/4		
Libertia		2–5										4/5
Ligularia			10/11							3/4		
Ligustrum			10/11		7–9	10/11					3/4	
Lilium	4/5	2/3	8/9						9–2			7/8
Limnanthes	4:9											
Limonium	4:7	2/3						1–3				
Linaria	3–5	2/3					4/5					
Linum	7/8	2–4					5–7					
Lippia				4–6								
Liquidambar			10/11									
Liriodendron			10/11									
Liriope										4:7		
Lithospermum					6–8							
Lobelia	4:7	2–4					8/9					
Lonicera	9/10				6–8	10/11					4:8	
Lunaria	5/6											

NAME	SEEDS			CUTTINGS						DIVISIONS		
	Cld	Htd	Cnd	Tip	S/M	Hdw	Bas	Rt	Bud	Div	Lay	Blb
Lupinus	6/7	1–4					3/4					
Luzula	7/8	3/4								4/5		
Lychnis	7	2–4					4/5					
Lysichiton	6/7									3/4		
Lysimachia				5	6–8		4/5			4:7		
Lythrum							4/5			4:8		
Macleaya		2–4					4	2/3		3/4		
Magnolia			8–10	4–6							4/5	
Mahonia	7/8								2/3			
Malcolmia	3–5											
Malope	3–5											
Malus	9–11											
Malva	4:7	2–4					4/5					
Matteucia										4/5		
Matthiola	7–9	3–5										
Mazus										4:9		
Meconopsis	3/4	2–5								3–5		
Melianthus		3–5			8/9							
Melissa										3:7		
Mentha										2–5		
Mertensia	7/8	2–4										
Mesembryanthe-mum	4/5	3/4										
Metasequoia					6/7	2						
Micromeria					6–8							
Milium	4:8	2–4								4/5		
Mimulus		2–5					4:7					
Mirabilis		2–4										
Miscanthus										4/5		
Moltkia					7/8							
Moluccella	2–4											
Monarda	7/8	2–4					4/5			4:8		
Morina	7/8	2–4										
Morisia								2/3				
Morus	10/11				7/8	11/12						
Muelenbeckia					7–9	10/11						
Muscari	6/7											7–9
Myosotis	7/8											
Myrrhis			9/10									
Myrtus					6–8	10						
Nandina			8/9		6–8	11/12			7			
Narcissus	6/7								7/8			8/9

NAME	SEEDS			CUTTINGS						DIVISIONS		
	Cld	Htd	Cnd	Tip	S/M	Hdw	Bas	Rt	Bud	Div	Lay	Blb
Nemesia		3–5										
Nemophila	4/5	3–5										
Nepeta		2–4					4/5					
Nerine		4/5										4–6
Nicandra		3/4										
Nicotiana		3/4										
Nierembergia		3–5										
Nigella	3:9											
Nomocharis		1–3	11–2						8–10			
Nymphaea							4/5			4–6		
Nyssa		2–4								10/11	9/10	
Ocimum		2–6										
Oenothera	4–7	2–4								4/5		
Olearia					10/11							
Omphalodes	7/8									7/8		
Onoclea		2–5										
Onopordum	5/6	4/5										
Onosma	2/3						7/8					
Ophiopogon			8/9							5:8		
Origanum							7/8			4:7		
Ornithogalum	7/8											6/7
Osmanthus					6/7						9/10	
Osmarea					6–8							
Osmunda	6–8									3/4		
Ourisia	4									4		
Oxalis										4/5		
Pachysandra										3:7		
Pancratium		2/3										8/9
Panicum	7/8	3/4										
Papaver	4:9	2–5						1/2				
Parahebe					6–8							
Parochetus	7/8	2–4								3/4		
Paronychia										4/5		
Parrotia											4:9	
Parthenocissus			9/10		6–8						4/5	
Passiflora					7					2/3		
Paulownia		2/3			7			12–2				
Pelargonium		11–2			4:9							
Pennisetum	7/8	3–5								4		
Penstemon	3/4	2–5			8/9							
Peonia			6–9							8/9	9–11	
Pernettya	2/3				7/8					4/5	9/10	

NAME	SEEDS			CUTTINGS						DIVISIONS		
	Cld	Htd	Cnd	Tip	S/M	Hdw	Bas	Rt	Bud	Div	Lay	Blb
Perovskia				4	7/8					4/5		
Petunia		3–5			8/9							
Phacelia	4:9											
Phalaris										4/5		
Philadelphus					6–8	10/11						
Phlomis	7	4/5			6/7	10/11				4/5		
Phlox		4/5			6–8		4/5	1–3		3/4		
Phormium		2–5								4/5		
Photinia					7/8	2						
Phygelius	4	2–4			6–8					3/4		
Phyllitus		3/4							4	4		
Phyllodoce					7/8					4/5	3/4	
Physalis	4:7									3/4		
Physocarpus					6/7	10/11						
Physostegia							3/4			3–5		
Phyteuma			12–2							3/4		
Phytolacca	7/8	2–4										
Picea	10–2				7/8							
Pieris	11–2				8/9						9/10	
Pinus	2/3											
Piptanthus		2–5			7/8						9/10	
Pittosporum	3/4	2–4			7–9							
Plantago	3–5									3/4		
Platanus	3	2				11/12					3/4	
Platycodon	7	1–4										
Polemonium	7	2–4										
Polygonum					6–9	10/11				4:8		
Polypodium										4/5		
Polystichum		4/5							9/10	3/4		
Poncirus		2–5										
Pontederia										4/5		
Populus						10–3				2/3		
Portulaca	3–5	3/4										
Potentilla	4:7	2/3			6–8	9–11	6/7			3–5		
Primula	1:7	2–5	12–2					1/2		3:7		
Prunella	3–5									3–5		
Prunus			9/10	4/5	7/8	2:10						
Pseudotsuga	1:10											
Ptelea	9/10											
Pterocephalus	3/4	3					4–6					
Pulmonaria	1/2									7/8		
Pulsatilla			7/8					7:2				

NAME	SEEDS			CUTTINGS						DIVISIONS		
	Cld	Htd	Cnd	Tip	S/M	Hdw	Bas	Rt	Bud	Div	Lay	Blb
Punica					6–8							
Puschkinia	6											6–9
Pyracantha			10/11		7–9							
Pyrethrum	7/8	2–4								4:7		
Quercus	10/11											
Ramonda			9/10						6/7			
Ranunculus		2–5										8/9
Raoulia							4/5			7–9		
Raphiolepis			10/11			9/10						
Reseda	4/5	2–4										
Rhazya		2–4										
Rheum	7	2/3								11–2		
Rhamnus	10/11				7/8	2						
Rhododendron	3–5	2/3			7/8	11:2					4:9	
Rhodohypoxis												9
Rhus		2–4			7/8			12–2		4	3/4	
Ribes					6–9	10/11						
Robinia	3/4	4								3/4		
Rodgersia		2–4								3/4		
Romneya		2–5						12–2				
Rosa			10/11		7/8	10/11		1/2	7/8	3/4		
Roscoea		2–4								4/5		
Rosmarinus					7–9	10–2						
Rubus					6/7	9/10				10–3	8	
Rudbeckia	7	2–4						2/3		4:8		
Ruscus			9/10							3/4		
Ruta	4/5	2–4			7/8	9/10						
Sagittaria										4/5		
Salix						11–3					4:9	
Salpiglossis		3–5										
Salvia	4:7	2–4		4/5	6–9	10–1	4/5			3/4	4	
Sambucus	2/3				7/8	1/2						
Santolina					6–9							
Saponaria	4:9									2–4		
Sarcococca			10/11		7/8	10/11				4/5		
Sarracenia		3/4								3		
Satureia		3/4			6–9							
Saxifraga	12/2			4/5			5–8			4:8		
Scabiosa	7/8	2–4					3/4			4:8		
Schisandra	10				8						9	
Schizanthus		3/4										
Schizophragma		2–4			8						4	

NAME	SEEDS			CUTTINGS							DIVISIONS		
	Cld	Htd	Cnd	Tip	S/M	Hdw	Bas	Rt	Bud	Div	Lay	Blb
Schizostylis										3–6		
Sciadopitys			10/11									
Scilla	6/7											6/7
Scrophularia							4/5			3/4		
Scutellaria	2:9									3/4		
Sedum	3/4	2–4					5/6			4:8		
Sempervivella										4/5		
Sempervivum		3/4								4:8		
Senecio	2–4				6–9	10/11				4/5		
Sequoia	3					9/10						
Sequoiadendron	3											
Shortia							6/7			5/6		
Sidalcea	4/5	2/3								3–5		
Silene	4:8	2–4					4/5					
Silybum	7/8											
Sisyrinchium	7/8	3/4								3–8		
Skimmia	9/10	2			7/8	9–11						
Smilacina										4:8		
Soldanella			12/1				5/6			6		
Solidago							4/5			4:8		
Sophora		3–5										
Sorbaria					7/8					10–3		
Sorbus			10/11									
Spartium	5	2–4										
Spiraea					6–8	10/11				3:10		
Stachys		3/4					6–8			4:8		
Stachyurus	10				7/8						3/4	
Stephanandra						10/11				3:10	4	
Sternbergia												7/8
Stipa										4/5		
Stokesia										4:7		
Stuartia			10/11		7/8						9/10	
Symphitum										4:7		
Symphoricarpos					7/8	10/11				10–3		
Symplocos			10/11		7/8							
Syringa				5/6	7					12–3	4	
Tagetes		3–5										
Tamarix						10/11						
Tanacetum					6–8					3/4		
Taxodium	10/11		2/3			11/12						
Taxus			10/11			9/10						
Tellima		3/4								4:8		

NAME	SEEDS			CUTTINGS						DIVISIONS		
	Cld	Htd	Cnd	Tip	S/M	Hdw	Bas	Rt	Bud	Div	Lay	Blb
Teucrium					6/7		5					
Thalictrum	2/3									4:7		
Thunbergia		3/4										
Thuya	2/3					4:9					9/10	
Thuyopsis		3/4				4:9						
Thymus					7/8		5/6			4:8		
Tiarella	3:7	2/3								4:9		
Tillia			10/11		6/7						9/10	
Tolmeia									6–9			
Trachelo-spermum				4/5	7/8						10	
Trachycarpus		4/5								4/5		
Tradescantia	3/4	2–5								4:8		
Trifolium	7/8									3–5		
Trillium	7/8											7–9
Trollius			7–9							4:8		
Tropaeolum	3/4	2–4						2/3		3/4		
Tsuga			11–2									
Tulipa	7/8											7–10
Typha										4/5		
Ulex		4			6/7	9–11						
Ulmus	6/7				8			12–2		2–4	9/10	
Ursinia		3–5										
Vaccinium	10/11			4/5		9/10				4	9	
Valeriana										3:7		
Venidium	5	3/4										
Veratrum			8/9							4/5		
Verbascum	4:7	2–4			6/7			2/3				
Verbena		2–5					4:9			3/4		
Veronica	4–6	2–4					7/8			3/4		
Viburnum			9–11	4/5	7–9	2:10					9/10	
Vinca					6–8	10/11				2–5		
Viola	4:7	1–5					7–9			7/8		
Viscum	2/3											
Vitis			11/12		7/8	11/12					3/4	
Weigela					6–8	10–12						
Wisteria					6/7						3/4	
Wulfenia	3/4	2–4										
Xeranthemum	4/5	3–5		–								
Yucca										3/4		
Zantedeschia												3/4
Zauschneria					6/7							
Zinnia	5	4/5										

General index

212

Index of Plant Names

214